THOUGHT X

Fictions and Hypotheticals

Edited by
Rob Appleby & Ra Page

First published in Great Britain in 2017 by Comma Press.
www.commapress.co.uk

A CIP catalogue record of this book is available from the British Library.

ISBN: 1905583605
ISBN-13: 978 190558607

This project has been developed through two separate grants from
the Institute of Physics in 2013 and 2015.

IOP Institute of Physics

The publisher gratefully acknowledges assistance from Arts Council England.

Printed and bound in England by CPI Group (UK) Ltd, Croydon CR0 4YY.

Contents

Introduction

THIS IS A BOOK ABOUT thought experiments in science and philosophy, and the way thought experiments have been used to poke holes in prevailing theories, or suggest limits to new ones. More specifically, it is about the stories these thought experiments tell – the unlikely scenarios they set up and the creative lengths they go to, often reaching blindly in the dark – to prove a hunch, that there's something wrong with our current thinking. The book will also explore how these peculiar 'short stories' of science and philosophy compare to their literary counterparts, what commonalities they share, and how they might illuminate each other.

So what is a thought experiment, exactly? The philosopher Andrew Irvine suggests the following definition:

> 'A thought experiment is an instance of reasoning which attempts to draw a conclusion about how the world either is or could be by positing some hypothetical, or perhaps even counterfactual, state of affairs.'[1]

The science writer Martin Cohen offers a more succinct phrase for this process – 'armchair philosophy'.[2] Indeed, we might expand this to 'armchair philosophy *and science*'. If the thought experiment is a successful one, by definition, there is no need to actually get out of one's armchair and conduct it in a real life laboratory or anywhere; the imagination's proof is enough.

The phrase 'thought experiment' (or 'Gedankenexperiment') was originally coined by the Austrian physicist Ernst Mach (1838-1916), who believed certain conclusions could be made without experiment, by all of us, by simply conceiving a scenario and then rejecting it for not complying with what he called our shared 'instinctive knowledge'.[3] Mach has a more poetic description of this 'instinctive knowledge' than anyone:

> 'Just as when the stroking of a clock has discontinued, we are still able to count the strokes in our memory; [Just as] we can still perceive in the afterimage of a lamp, details which escaped immediate observation; we can, in recollecting, discover a feature which suddenly unveils for us the previously misunderstood character of a person. Similarly, in our recollection, we can discover new properties about physical facts which for a long time went unnoticed.'[4]

It doesn't sound very scientific, does it? Merely thinking about a theory can, according to Mach, lead to new insights and new developments in our understanding of the world. Sometimes, perhaps, that's all scientists *can* do. As Ernest Rutherford once quipped, 'We've got no money, so we've got to think.'[5]

To understand how a thought experiment is constructed, let's take one of the earliest known examples: the Leaning Tower of Pisa thought experiment. Devised by Galileo, this took the existing belief (proposed by Aristotle) that heavy things fall faster than light things, and proved this theory to be false, everywhere, in principle. He did this by imagining an experiment that involved dropping things from the Leaning Tower of Pisa (an experiment which, as far as we know, he never actually conducted). If heavy things fall faster, then imagine tying a heavy object, let's say a cannonball, to a light object, let's say a musket ball, with a string. Drop them both together. Considering the cannonball, musket ball, and string as one system, the system is heavier than just the cannonball, so as a whole, according to Aristotle, it should fall faster than just a cannonball. But consider each constituent part individually. The musket ball, according to Aristotle, is inclined to fall slower than the cannonball, so it would drag the cannonball back, slowing it down, thus making the system as a whole fall slower. So we have a paradox. Both can't be true, so Aristotle's theory can be rejected as false.

This 'deductive' thought experiment, is the strongest type. It sets up a reasonable situation where the original theory produces a logical paradox, in principle, everywhere. Something can't be true and not true, so the original theory has to be rejected. (Interestingly, Galileo conceived this argument to counter the many occasions in nature where lighter things *do* fall slower than heavier ones – a feather compared to a stone, let's say – because of what we now know is air resistance. To distinguish the two, Galileo had to divide all possible causes into two types: fundamental 'phenomena', or 'laws' as Newton would later call them; and 'accidents', by which he meant interference factors that get in the way of seeing these fundamental laws – a distinction that at the time took an enormous leap of faith).

There is a second, softer type of thought experiment, examples of which merely try to demonstrate something is missing in the current theory. We might call these 'limiting' thought experiments. The famous Twin Paradox, explored by the first story in this book (and which ironically *isn't* a paradox), belongs in this category. It points out a lack of symmetry in Special Relativity, producing an intuitive sense that something is missing. If the timeframe inside a very fast moving spaceship appears to be slowed down (relative to us), when it's moving away from us, and sped up (relative to us) when it's coming back and moving towards us, then surely everything will even out, for the travelling twin, in the thought experiment, who performs a 'round trip' there and back?

As Adam Marek's story 'Lightspeed' (pp.1–17) demonstrates, this isn't the case; the travelling character doesn't age as much as those he left behind, time has definitely moved slower for him. So where is the asymmetry coming from? Why can't we say, 'The stationary family are also *moving* – away and then back again – relative to the travelling pilot'? An asymmetry is needed here, and to introduce it (and fully explain it) Einstein had to develop a second theory, General Relativity, which included the all important component, *acceleration* – the one property of the travelling twin's timeframe that distinguishes it from the stationary twin's.

What the Twin Paradox does here is show that there is a 'back wall' to Special Relativity, a point where a new theory, or at least an extension to the existing one, needs to be built.

Another example of a 'limiting' thought experiment, although not a successful one, is Maxwell's Demon (pp.97-112). The Second Law of Thermodynamics says, among other things, that if you put a hot cloud of gas beside a cold one, heat will travel from the hot to the cold; heat will eventually 'even out'. The Maxwell's Demon thought experiment speculates about the possibility of a tiny monster (or nanobot) opening and shutting a tiny door in a wall between a hot and cold gas, only letting colder-than-average particles in the hot gas, and hotter-than-average particles in the cold gas pass to the other side, thus making heat move in the *wrong direction*. Here we see the creative mind of the scientist in full flow, inventing very elaborate and unlikely scenarios, simply to deliver pot-shots at an existing theory. What drives the scientists behind these ideas to such imaginative lengths? Professional envy? Innate belligerence? Or perhaps it's just the same dogged curiosity, the impulse to test, test and test again, that drives all scientists in all aspects of their work. Even in the absence of technology or funding to do 'real world' experimentation, test they still must, if only in the mind.

There is perhaps a third type of thought experiment – let's call it the 'intuitional' thought experiment – that doesn't condemn a theory altogether, or call for an extension, but raises the suspicion that it's *conceptually* incomplete. A very famous example of this is Schrödinger's Cat (pp.139-158). Quantum Mechanics says that, at a microscopic level, things behave very strangely; subatomic particles, like electrons, can exist in a 'superposition' of different states at the same time – a probabilistic amalgam (or 'wavefunction') of 'spinning one way' and *simultaneously* 'spinning the other way', for example.

It's only when we measure them, says QM, that the 'wavefunction' collapses into one state of affairs or another; only when we measure it does it 'decide' what it wants to be. Weird, right? Certain naysayers in the early days of QM suggested the Cat thought experiment as a way of extrapolating this strangeness to a macroscopic, human level. 'Are we really saying that the cat in the unopened box is both alive *and* dead, before we open it?' they asked. This thought experiment isn't a disproof, exactly, but it is an appeal to our intuition, our 'instinctive knowledge', that this *can't* be what's going on? It's interesting to note the way the argument cleverly transplants the problem into a very everyday context. As Mach said, instinctive knowledge's 'only value is in the provinces with which we are very familiar'.[6] You can't get much more familiar than a household pet.

Frank Jackson's 'Mary's Room' thought experiment (pp.65-80), also known as 'Mary, the Colour Scientist', is another example of this 'something's missing' hunch. It suggests there's something awry in the physicalist's argument that 'all knowledge can be represented by statements about what is physical'. It knows there's something wrong with this theory, but it can't quite put its finger on it. When the woman in the scenario sees colour for the first time in her life (after previously living in a entirely black and white room), the physicalist argues that she's learning nothing new, nothing that she couldn't have previously learned in the room by studying the science of colour (and other scientific facts). This is not a disproof by deduction, or a limitation of the theory to certain conditions, it's an appeal to our intuition. Something's wrong here, it says. Surely Mary learns *something* new the first time she sees the colour red?

So, it seems, we have an assortment of different *types* of thought experiments with different features.

It was interesting, putting this project together, to see how strongly some people felt about what qualifies as a thought experiment. Professor John Norton, in correspondence, declared that the Unexpected Hanging, for instance, was 'not a thought experiment but a pernicious logical trap!' Professor Thomas Nagel, when asked about his Spider in a Urinal argument (pp.219-222), insisted this too 'isn't a thought experiment but a true story.'

A theory has to be applied; it can't just be a logical puzzle or word game. And it must be fiction (to use Andrew Irvine's terms, not just 'hypothetical' but 'counterfactual'). It needs to tell a story, indeed it needs to do more than that it needs to create a 'story-world': one in which the theory is not just challenged but explored and communicated, in which different interpretations are possible; and into which (as a fictional space) all-comers are welcomed – not just tenured academics and professional scientists, but everyone. Everyone has a contribution to make to the discussion; there are plenty of armchairs to go round! It's only by playing out the 'story of a theory' that we can see if some unexpected essence or undiscovered consequence emerges. In this sense, a thought experiment is like the first time a new computer game is played, by queuing fans or in-house testers. It's one thing to design such a computer game (i.e. invent a theory), but it's another to see exactly where playing that game might take us.

That's why, for this anthology, we haven't just let scientists 'play the game' of a thought experiment, we've invited authors to play it too. If a thought experiment is just, at base, a story like any other, then it too can be mirrored by, or woven into, other fictional stories. That's the hypothesis we're testing in this anthology. Here, 14 authors have tried to do just this, working with scientists and philosophers for whom the thought experiment in question is very dear (in one case, the consultant is the *inventor* of the thought experiment), to produce stories that 'play out the game' of a theory in a new way. From Galileo and Einstein's parallel adventures into relativity, more than 300 years apart, to John Searle and Frank Jackson's interrogations of consciousness, to Robert Nozick's speculations on the relative merits of happiness – both physical and philosophical questions are explored.

Ernst Mach was always keen to highlight 'the identity [or the similarity] of the creative imagination of the artist and the scientist'. Both, he claimed, deploy a 'continuous change of visual imagination', going through various iterations of a scenario with different starting points, until they alight on their 'story' or 'scene'. Mach also starts off his key essay, *On Thought Experiments*, by defining 'experimentation' as simply 'the method of variation'. Literary artists (i.e. writers) do just this – scroll through an imaginary rolodex of different scenarios to find the right one. So it only seems right that they and scientists should be brought together to share, explore and celebrate these inventions.

Perhaps there's one last way in which thought experiments are like stories. Consider Mach's original, very poetic comparison between what a thought experiment (based on 'instinctive knowledge') does, and the experience of discovering some feature of a 'previously misunderstood character'. Isn't this exactly what a good short story does? Isn't this delayed revelation of something which 'for a long time went unnoticed' exactly what an epiphany in a short story is? It makes us re-evaluate something we should have seen before, something we had the means to work out ourselves all along. It might not be possible to paraphrase exactly what this epiphany is, but it taps into something that's common to all, a shared intuition.[7]

It's our hope that these stories, like the thought experiments that inspired them, will produce similar revelations.

Rob Appleby & Ra Page

Notes

1. A. D. Irvine, 'Thought Experiments in Scientific Reasoning', in T. Horowitz, and G. Massey (eds), *Thought Experiments in Science and Philosophy* (Lanham: Roman and Littlefield, 1991), pp.149-165.

2. M. Cohen. *Wittgenstein's Beetle and Other Classic Thought Experiments* (Oxford: Blackwell, 2005), p1.

3. The example Mach first applied the idea to was a thought experiment suggested by Simon Stevin or Stevinus (1548-1620), involving a perfectly smooth, pulley of triangular-cross-section, and a circular chain of equally-spaced, equal weights draped over it. Stevinus began by discounting a very real scenario – that the chain rotates around the pulley continually. As the pulley's surfaces are perfectly smooth, and weight distribution equal, this is a genuine mechanical possibility, but the intuition discounts it. Perpetual motion machines don't exist, intuition tells us, so this can't be a possibility. *The Science of Mechanics: A Critical and Historical Account of its Development* (1883), translated by Thomas J. McCormack, (London/Chicago: Open Court, 1893, repr. 1919), pp.24-30.

4. Ernst Mach, 'On Thought Experiments' (1897), translated W.O. Price and W. Krimsky, in *Philosophical Forum*, 4, 1973, pp446-457. 'Über Gedankenexperimente' by E. Mach, from *Erkenntnis und Irrtum: Skizzen zur Psychologie der Forschung*, (Leipzig, repr. 1906), pp.181-197.

5. R. V. Jones in the *Bulletin of the Institute of Physics* (1962) recalling the dictum of Ernest Rutherford.

6. Ernst Mach, 1883. p27.

7. Incidentally Mach later clarified and perhaps tempered his idea of 'instinctive knowledge', in the 1897 essay, saying what's really instinctive or innate is our impulse to experiment.

Lightspeed

Adam Marek

'IT'S HIS JOB,' MARTHA said.

'Let's try not to make definitive statements. Remember, there is no objective reality, only perceptions.'

'Right.' Martha rolled her eyes. 'To me it *feels* like it's his job.'

'What specifically is it about Nowak's job that feels like a problem to you?'

'He's away so much.'

'That's how it appears to you, anyway.' Nowak said.

There was a loud clunk, and a grinding noise, and then the inertia warning lamps ignited on the ceiling. They all fastened their seatbelts.

'That's not just my perception. It's your daughter's too.'

'Don't say "my daughter" like I'm not...'

'Let's try not to issue commandments,' the counsellor said.

'Fine. When you say "my daughter" it feels like you're accusing me of not being a good father, which of course I am. I'd be grateful if you could just use the name we gave her.' He looked to the counsellor for reassurance that he'd phrased it within the bounds of their session agreements. She gave him none.

The station's rotation slowed and Nowak felt his weight diminish. Their leather seats creaked as their buttocks lifted away from them. Martha squeezed the arm of her chair.

'I feel sick,' she said.

'You're... I mean, it appears to me that you are making *me* the fulcrum of all the dissatisfactions in your life, many of which have nothing to do with me.'

'The station, hon. I meant the station.'

'Oh.'

The following morning, the station ring was still failing regularly, the rotation slowing and the verisimilitude of gravity falling away, making the contents of coffee cups pitch, and then all the objects of their life to become animate pioneers of the air, suddenly blessed with a thirst for exploring the interior of their pod.

A blob of coffee the size of a tennis ball collided with Martha's chest, the impact causing gelatinous ripples to whirl across its surface, warping its shape into a mesmerising landscape of waveforms, without breaking the ball's integrity. It rolled up and over her shoulder towards the ceiling, where it was caught in the suck of a vent and whipped out of their lives forever.

'For fuck's sake,' she said.

Gretchen foraged in the air for her cereal, like a fish, hoovering up pebbles of milk with noisy slurps.

'Put your seatbelt back on,' Martha said.

'Don't poo-poo her ability to find the fun in this,' Nowak said. 'We don't all have to live in Irritable-City,' and then to Gretchen, 'do we sweetheart?' He undid his own belt, and pushed off from his seat to retrieve his toast.

'Don't make her your ally against me.'

'Hon, *please* chill out. Try to find *some* joy in it.'

The habitat ring of the station made a metallic groan, like an iron giant stretching its limbs, and then the rotation resumed, the objects in the room slowly acquiring weight.

When they had finished breakfast, Nowak put the dishes in the washer, and then allowed Gretchen to prepare their home for the next phase of the day. To her, the pod was a wonderland, the touch of her chubby little fingers causing the mutable framework to reconfigure with gentle pneumatic puffs. The beds glided, swan-like, back inside the wall, and the dinner table spiralled down into the floor like an enormous white sycamore seed. Their shower cubicle was whisked away, and their lounge delivered.

'Did you book the Zero Grav Suite for Gretch's birthday?' Martha said.

'Of course.'

'I've only had two acceptances so far.'

'We live within a hundred metres of everyone we invited. No one's going to miss it.'

'It's less than a week away.'

'Everyone will come. I promise.'

'I just can't stand this.'

'There is no *this*. It's all this,' he tapped the side of his head.

'Don't patronise me.'

Nowak squeezed her shoulder.

'I have to go to work.'

'Right.'

'I'll see you later.'

'Later for you, but not for me.'

Nowak sighed and moved to kiss her, but she turned away.

'Let's just hope I make it back alive then and this isn't your final memory of me.'

'Great. Thanks for that.'

The anaesthetist checked Nowak's name, which was a ridiculous bureaucratic necessity, as Nowak was one of only three lightspeed pilots, and he the only male one.

'I've lowered the dose a little this time,' she said as she pulled down the collar of his wet suit and strapped the air-cannula to his neck. 'When you came out of lightspeed last time, your responsiveness was a little off.'

'It was?'

As far as he was concerned, he was fully alert throughout the voyage. He'd made the manual adjustments to the ship's trajectory, arrived and docked with Leapfrog-One to within a few minutes of the scheduled time. And when you considered the distance he was travelling, and the time-warping effects of travelling at 0.94 of the speed of light, this was an incredible feat, a level of accuracy that neither of the other two pilots was capable of matching. So the anaesthetist's implication felt unfair.

Nowak put on his helmet and climbed into the cockpit of the Lightship, a gleaming one-person bullet that both excited and terrified him. This was *his* ship, its interior built specifically for his body. He pushed his arms and legs into the soft compartments, felt the guts of the ship activate, shrinking to fit his body in a tight embrace and immobilising his limbs.

'All set,' he said.

The anaesthetist closed the hatch. Nowak gave the command, and the fluids began to fill up the tiny space, cocooning his body in warm amniotic goo. The ship made a sucking noise, then a sigh, increasing the internal pressure till he was held in its rigid suspension as firmly as a prehistoric wasp in amber. He felt the kiss of the air cannula activating on his neck. What was Martha doing right now? Was she thinking about him? How about Gretchen? Had they done the right thing? And then he was ejected softly into space, the GABA-Glycine flooding his system, severing his body's connections with his brain, incrementally paralysing him from the neck down till he felt like he was just a head in a jar as the monstrous acceleration began.

When Nowak had first been offered the job, Martha had seemed even more excited than he was about living off-world. She spent more time on the phone that week than she'd spent on it for the previous year, saying delighted and emphatic '*I know!*'s to her friends and family members when she delivered the news. These people had many questions about the practicalities of life on the station, about what they'd do with their home in Florida, their stuff, how they expected to maintain friendships for four years through messaging and video calls alone. These were things Nowak and Martha had already talked about on the day he'd been offered the job, before they made the actual decision and he'd accepted.

Martha's answers to her interlocutors were always reassurances, reframing space emigration as being – for all intents and purposes – not that dissimilar to moving to New Zealand, something one of her sisters had done a few years ago, without any noticeable detriment to family cohesion.

And when they asked about her work – what would she do up there? – she'd said that the opportunity to move her studies into space was something her company were very excited about. Investigating sperm motility in microgravity might provide information about infertility that could lead to pioneering new treatments on Earth.

But aside from the practical questions about moving, working and maintaining relationships, there had been another level of questions for them to consider: about taking an eight-year-old child into space, a change which had no analogue on Earth. How would the imperfectly artificial gravity and the imperfectly screened solar radiation affect her development? Would her brain be able to develop connections in the way that the brain of a normal Earth child would? And how about the exposure to cosmic rays, some of which might still evade the filters in the station's fuselage and tear through its personnel like lightning bolts, leaving long fulgurites of mutated DNA through their bodies? Would deletions and aberrations in her genetic code be more significant in her developing body than in the body of someone who came to the station as an adult? After four years of pre-puberty growth in space, would her bones become irrecoverably fragile so that the Earth's gravity became a lifelong danger to her?

These questions had been debated fiercely in the media six years before, when the first child had moved up to the station: a four-year old boy named Luka who was the son of the world's foremost astro-agronomist. He'd set the precedent, and since then, many other children had spent significant time in space. Gretchen would be the twenty-third child to live on the station, and one of fifteen currently there.

There were too many unknowns for there to be a 'right'

decision, and any reassurances the company offered about the safety of the station for a child were loaded with caveats. In the end, he and Martha had agreed that yes it would be a risk, but that risk would be offset by the amazing education she would be able to receive when they returned to Earth, thanks to Nowak's new income, and by the enormous benefits to her career prospects from having spent significant time off-world.

In the end, Martha asked the most pertinent question, which framed the whole situation in a way that allowed them to make the decision and stick to it without guilt: would Gretchen thank us for it when she was older? They believed she would.

Through the internal speakers of Nowak's helmet came birdsong. Percussive ticks and mellifluous warbles, each one accompanied by the voice of an English man relating the name of the bird: *chiff-chaff, black-cap, wood-warbler*. There were 300 birds in all, and having made 30 such trips, Nowak was already able to recognise more than half of them, saying the name in his head before the man announced it. What a thing it must have been to walk in the woods and hear this symphony, all these hundreds of birds singing at once.

It was important for his brain to have a focus point. The journey to Leapfrog-One was a two-hour brain-yawn, the dilatory effect on time of travelling at this velocity causing thoughts to yo-yo between moments of intense focus, where a single pulse in the machine-gun trill of a mistle thrush was opened up into an aural desert of intense meditative complexity, and moments where his focus slipped, and he tumbled down the cacophonous staircase of a dozen bird calls in a blink.

His mind's struggle to maintain continuity of processing speed was most intense during the acceleration phase. For the portion of the journey when he clipped through space at a constant 282 million metres per second, the effects of this time-sickness were greatly diminished. And with the ship moving at a constant velocity where wavelengths once again became regular, its computer was able to establish a signal with the station.

He commanded the ship to call home, and Martha picked up.

‘IIII’MMMMMMM BBBBUUUUUUSSSSSYYYYYY,’ she said, her voice coming to him at about a third of its normal speed. She was chopping carrots, her usual staccato rhythm drawn out, so that she appeared to be paring each fresh slice with supreme tenderness. She didn’t like speaking to him like this, the deepness of his slow-voice for her creating an uncanny effect, the stuff of nightmares. But Gretchen loved it. He asked Martha to put her on.

‘Hhhhheeeeeeeeeeeellllllllllllloooooooooooooo ddddddaaaaaaaaaaddddddddyyyyyyyy.’

Nowak laughed. Gretchen thought that because his voice came to her so slowed down, she needed to slow her own voice down and deepen it to communicate with him, and this exaggerated the time dilation effect even more.

‘How was your day?’ He’d only left the house a couple of hours ago, but on the station, over 9 hours would have passed for Gretchen. He could see she was already in her pyjamas.

‘Iiiiiiiiiiiitttttttttttt wwwwwwwaaaaaaaaaaaaaasssssssssss fffffffffffiiiiiiiiiiiinnnnnneeeeee tttthhhhhhaaaaaaaannnnnnnkkkkk yyyyyoooooouuuuuu.’

It was difficult to have a proper conversation like this, but it was important to him that she knew he was thinking about her, when for her he was away for so long. They exchanged a few sentences, in which Gretchen explained that she’d been to the station’s farm and planted rice today. Nowak switched the visual feed to the Lightship’s external camera, so she could see the neon-spaghetti sky, a marvel that made her smile every time.

‘I love you sweetheart. I’ll see you soon.’

‘Iiiiiii llllllooooooooovvvvvvvvveeeeeeeeee yyyyyyoooooouuuuuu tttttoooooooo.’

After the anaesthetist’s comments on his performance last time, he was determined to arrive as close as possible to the scheduled time. He issued commands, correcting his trajectory and velocity as the stars shrank from luminous strands to pinpricks, watching the difference between the scheduled and estimated arrival time diminish.

And then Leapfrog-One came into sight, a 60-person titanium limpet clamped to the side of asteroid ZFG36. Making its way out to a position half a light year from Earth, this piloted-asteroid would eventually become the first of four intergalactic pitstops, perhaps paving the way for the future colonisation of exoplanets. The big stuff.

Docking commands began to appear inside Nowak's helmet as Leapfrog-One paired with the Lightship. There was nothing for him to do now but wait. Everything from this stage was automated, Leapfrog-One and the Lightship docking to do a quick plutonium handshake, swapping the depleted cores on the Leapfrog for the fresh ones stashed in the hold of Nowak's ship. The whole process would take 90 minutes, meaning Nowak could just lie like a bug in his hyperspace pupa and enjoy the view: sunrise over the asteroid's pumice horizon, the fireball's volatile corona visible through the ship's filters at this distance, and Saturn in transit. This was the environment in which he was bringing up his child. He hoped that when she was an adult, looking back on her childhood, she would appreciate it. He took a shot and sent it back to her with the message:

Gretchen, my angel, you make this look boring.

'I'm on my way home,' Nowak said.

Gretchen laughed at the sound of his voice, which to her, now that he was flying at close to light speed *towards* her, would be sped up and mouse-like.

'Hlo ddy,' Gretchen said, all her fidgeting now delivered in rapid bursts. In the background, Martha whizzed about their pod tidying up in a comedic frenzy.

'I'll be back in a couple of hours. Are you excited about your party?'

A look of puzzlement blinked across her face.

'T er ysdy.'

'What was that my love?'

Martha said something to Gretchen and the girl shifted off camera, making way for her mum. Martha's face appeared, large and angry, in the screen. Her eyebrows pinged up high, her lips zipped into a tight knot.

'Yu mssd t,' she squeaked.

'What do you mean?'

'T's Sndy Hr.'

She stared at him.

'That's impossible. It's Friday.'

'T's Sndy,' she said, and then her hand flew up past the screen and the connection was severed.

'I'm not sure how that's even possible,' Nowak said.

'This is the problem, hon. Your reality is very different to mine and Gretch's.'

'Why don't you tell Nowak how it is for you?' the counsellor said.

'He knows how it is but he doesn't care.'

'You're mind-reading,' Nowak said, using one of the counsellor's terms. 'You're making assumptions about how I feel. I don't know how you can think I don't care. You've no evidence to support that.'

'I'm just stating the facts, from mine and Gretch's point of view.'

'Gretchen isn't here to state how it feels for her so you're only qualified to talk for yourself.'

'Oh for goodness' sake. Now you're being evasive.'

'Martha, you were telling Nowak how it feels for you.'

'It feels like me and Gretchen made a huge sacrifice to our quality of life for your career.'

'What about your work?'

'You never even ask about it.'

'Well I know it's been disappointing for you and I didn't

want to upset you by asking you to talk about it. And you know very well we didn't come up here just for...'

'Don't kid yourself hon. It's the only reason we...'

'That's not true.'

'And then you have the gall to complain about your work when you get back, so I sit here waiting and not knowing when you're coming back and thinking why the hell did we make this move if none of us is happy?'

Nowak thought about this for a moment.

'Okay, so, to be honest...'

'You don't need to preface your statements with that phrase,' the counsellor interjected. 'The assumption is that everything you say here is honest, and by stating that, you destabilise Martha's ability to trust you.'

'Jesus christ.' And then to Martha, 'I sometimes think we'd solve this much quicker by ourselves at home by throwing a few dishes at each other. This is ridiculous.'

Both women stared at him.

'Fine. I will now explain why I behave that way sometimes using the same terms of truthfulness which I have displayed throughout our conversations in this room, and all our rooms for that matter. I am an honest man and I have nothing to hide from you.'

He raised his eyebrows, but neither Martha nor the counsellor reacted in any way.

'I'm sorry that I moan about my work when I come back. The truth is... just let me carry the fuck on will you?... the truth is that I love my damn job. It's amazing. I'm a space pilot and not just that, I'm a space pilot for the most advanced craft ever, and not just that but there are only three people who are skilled enough to do it and I'm the best of them. When it comes to pilots, I'm a fucking genius. I'm the best on Earth. And off it. I get to see things that no one has *ever* seen and travel at speeds that are barely within the threshold of survivable. But when I come back, and you're moping about, and the time screw-ups mean I mess up and miss Martha's party or whatever the hell else it is that that I've done, I moan about it because I can see that you hate it up here and if I'm enjoying it then it will only make the crap time you're having seem even more crap, relatively speaking. I thought if you thought I was having a crap time too then it would be easier for you.'

'Well that's ridiculous.'

'Please understand that, however misguided or ineffectual, all my behaviours are coming from a positive intent, that being your and Gretchen's happiness. I'm sorry if your job up here has turned out to be dissatisfying, but we had no way to know that would happen and I'm powerless to do anything about it but try to make you feel better.'

The counsellor nodded. 'And what are your *facts*, Martha?'

'My facts are that I never know when Nowak...'

'Talk to Nowak, not to me.'

'...I never know when you're coming back. You call and say you'll be back in a few hours, but then you don't show up for two days. And it's only going to get worse because you're travelling further away each time. What's it going to be like in a couple of years?'

'You know there's nothing that I can...'

'Yes but put yourself in my position. And it's even harder for Gretch.'

'I don't like to suggest solutions,' the counsellor said, 'but, can I ask...I understand with the time warping effects...' she swirled her hands around in the air as if travelling in a Lightship was like tumbling down a plughole. '...Everything gets messed up and confusing. But is it impossible to predict the time you'll be back or just difficult?'

'It's tricky because any minor course adjustments at those speeds might delay me by hours, or longer. But the guys in Control are making those calculations while I'm en route... so I guess I could get them to keep you updated?'

'How would that be, Martha?'

'That would be a help, yes.'

The counsellor patted her thighs. 'Good,' she said, apparently enjoying this accomplishment. 'I'll see you both next week.'

Control sat at the top of the central spine of the station, a part that did not rotate, and so the personnel here were strapped to their chairs, tracking and co-ordinating the movements of more than three dozen spacecraft ferrying between the Earth and the station. On the ceiling, five people were floating at a control panel, hanging onto the handles around it, and desperately trying to fix the habitat ring, which had been jammed now for three hours.

'Between you and me,' Shiva said, when Nowak asked, 'it appears there's some old redundant emergency protocol built deep into the station's software that's being triggered somehow, but no one can work out why.'

'How much longer's it going to go on?'

Shiva shrugged. 'Worst case scenario, they'll have to rebuild the operational software from the ground up. It could take a year.'

'What the hell?'

'But they'll find a temporary work-around.'

'Right.'

'So what I've done…' Shiva said, looking excited and handing him the data, 'is calculate your arrival times, relative to the station, within a 12-hour window of error, based on your previous trips.'

'This is great, thank…'

'And not just that. This is actually very interesting, but I calculated your journeys for the entire rest of your posting, compensating for the increasing distances. The latitude for error obviously gets wider every time, so by year four, it's only possible to predict your arrival to within 36 hours either way, but of course, that's just from this point in time. At the time of the actual flight we'll narrow that down significantly.'

Nowak looked at the data, and could see that in year four, a journey that for him would take 19 hours, would mean he was away from Martha and Gretchen for at least 12 days. He could imagine Martha's reaction to *that* information.

'Thanks mate,' Nowak said.

'No, thank you. This is just the kind of challenge I enjoy. I actually calculated, and you'll enjoy this, that even though you're contracted for four years, that's four years relative to this station. In your reality, you'll actually only put in one year, four months, and 17 days. So your pro-rata income is actually almost triple what you're on. Pretty cool, huh?'

One part of Nowak's brain was relieved that this situation, which was putting such a strain on family life, would be over much quicker than expected, but of course it would only be that way for him. He'd have to keep this perspective on things from Martha for as long as possible.

'So…hang on. Does that mean that over the next four years, I'll only age a year and a bit?'

'Yep.'

'So by the time we all return to Earth, I'll only be 35, but…'

'Well, your body will only age a year and a bit, but I'm not sure you could get everyone else to adopt your own personal biological calendar. I guess what'll actually happen is that you'll have a birthday every few months. Which I suppose will make you feel much older in fact.'

'Great.'

'But, in terms of your body, yes, you'll only age a year and four months and 17 days. Which means you might live to be three years older than you would have done before.'

'So… at the end of the contract, my body will actually be younger than Martha's? She'll overtake me?'

Shiva looked up at the ceiling, thinking about this. 'Yes. Hah. You'll be married to an older woman.'

'Jesus, this was supposed to make her feel better.'

'Glad to help.

Nowak watched Martha scrolling through the information. Behind her, in the window out to space, the Earth came into view: nine billion dramas locked inside a marble.

'This is depressing reading,' she said.

'I didn't think it was so bad?' He'd only given her the next six months of scheduled flights. While Nowak valued openness and honesty as the core of a stable and happy relationship, it was better that Martha discovered the facts slowly, had time to adjust to the idea. If he fed her the details piecemeal, their marriage might actually survive the experience, and when they returned to Earth, her memory would shrink all those days and weeks of waiting and not knowing into single blips of time, stripped of their raw emotions. They would be happy again and enjoying all the benefits to their lives that had motivated them to come up here in the first place. Everything would be good and she'd be glad of the sacrifices she'd had to make.

'It's only three and a bit years,' he said. 'Three years ago, we were just moving to Tampa. That seems like yesterday…'

'Hah, right.'

'In a way.'

'I'm going crazy up here. I don't know how much more I…'

Nowak put his hand on her knee. 'You're the smartest person I know. You've just got to find the resources to get through this. Don't think about it as being trapped up here without me half the time, think about…'

'It's going to be more than half the time.'

'Focus on the possibilities. There's so much potential for you and Gretch up here.'

'That's just not true. Since my funding got cut, I've got a couple of hours' lab time a week. That's not enough to achieve anything.'

'So submit another proposal.'

'That could take years to come through.'

'Not necessarily.'

The door to the pod opened and Gretchen came in with her school bag.

'Honey?' Martha said.

'They closed school early.'

'How come?'

She shrugged.

'They just did.'

Martha sighed and handed the data back to Nowak. 'We'll talk about this more another time.'

Nowak felt sick. Was Martha also drip-feeding him, getting him used to the idea that she was going to leave with Gretchen and return to Earth? He couldn't be without them, but he was stuck here. His contract was strict, and the penalties for early severance serious. He would never be able to forgive Martha if she forced him to cut short the best gig he'd ever have in his life.

The PA system made a double ping sound, and then an announcement was made. 'Would all station personnel assemble in the habitat ring corridor immediately.'

Martha looked at Nowak, concerned. 'Something's not right.'

The habitat ring was the only place large enough on the station to accommodate all 700 personnel. As they left their pods, exchanging worries about what was happening, the looped chamber did something acoustically freaky to their chatter. There were moments when the voice pitches of particular speakers became numerous and loud enough that they set the air ringing like a tuning fork, these hums causing Nowak's whole body to tingle. Gretchen had her hands over her ears.

'It'll be over soon sweetheart,' he said.

Emergency situation wardens in bright orange tabards moved through each segment of the habitat ring taking a register. And only when everyone was accounted for did the announcement begin. It was Amrir Cortez, the Station Commander.

'Thank you for your patience. As you are aware, we have been working to overcome the recent issues with the habitat ring rotation. We believe we've isolated the cause and have identified a solution, which will only take six-to-eight hours to implement.'

The chatter resumed with sounds of widespread relief.

'The work-around requires a full-system reboot, meaning we'll have to shut down power temporarily to the whole station. This includes all life-support systems.'

There was a loud pinging sound, a tone repeated five times, and then the door-locks of every pod clicked, the indicator lights turning from blue to red. Concern now made the timbre of the chatter rise in volume and pitch, the harmonics thrumming in Nowak's ears.

'This is bad,' Martha said.

'Jesus, would you…' Nowak picked up Gretchen and held her against his chest. She wrapped her legs round him and rested her head on his shoulder. 'Everything's going to be okay sweetheart,' he said and kissed her ear.

'Please be calm,' the announcement continued. 'There is nothing to be concerned about. You are all safe. Please will you all now move calmly to the lifeboats.'

The ring juddered to a halt and everyone was cast up into the air, their limbs entangling. A foot struck the back of Nowak's head. He raised his knees to curl himself protectively around Gretchen, and said again, 'Everything's going to be okay.' The lights blinked out, and for a moment the corridor became a clamorous whirlpool of bodies, before they blinked back on, the ring's rotation resumed, and everyone descended gently to the ground, picking themselves up and complaining vociferously to the emergency wardens.

'That wasn't part of the evacuation,' one of the wardens said, and the word evacuation whipped round the room, escalating the panic further.

'We shouldn't be here,' Martha said. 'We made a mistake. A big mistake.' Her gaze was darting everywhere, frantic.

'Hey,' he said, stroking the side of her face. 'We're going be fine. It's just a blip.'

'Please be calm,' the Commander said. 'Apologies for that. Will you all now make your way to the lifeboats. Once all station personnel are on board, we will uncouple from the station and move to a safe distance before beginning the station reboot. You will be away for just a few hours, and then we have every confidence that these problems will be resolved and you can resume your normal routines.'

'There you go,' Nowak said.

'Can't you see what's happening? Look where you've put us!'

The airlocks to the tunnels which led to the central hub of the station opened and the wardens struggled to maintain order as people jostled and shoved their way forwards.

'This is not an emergency situation,' the Station Commander announced. 'Please move with courtesy. There is no danger other than the danger you create yourselves by panicking.'

'Are we going on the lifeboat?' Gretchen said.

'Yes honey,' Martha said, 'We're getting out of here.'

'It's an adventure, sweetheart,' Nowak said, concealing his own fears to give Gretchen his most reassuring smile. 'You'll be able to see the whole station from out there. It's going to be amazing and you'll remember it...' he rubbed the tip of his nose against hers, '...even when you're a wrinkly old lady.'

Afterword:

The Twin Paradox

Prof. Tara Shears

University of Liverpool

ADAM'S STORY REVOLVES AROUND the strained relationship between Nowak, a space pilot, his wife and their young daughter, who are stuck at home in an inflatable space pod during Nowak's long missions. The tension between them isn't helped by the peculiar behaviour induced by the laws of high speed physics. Nowak's space journeys are subject to the laws of Einstein's theories of relativity, and that makes his life unpredictable, sometimes willfully so in the eyes of his wife.

Understanding how things move at high speeds should be simple. We have an intuitive idea of space and distance and velocity – we know that it will take us an hour to walk four kilometres if our walking speed is four kilometres an hour, and half an hour if we speed up to eight kilometres an hour. We have an intuitive idea of time and how long a minute and a second last. But that grasp of space and time that serves us so well normally causes no end of problems when it comes to thinking about anything travelling near to the speed of light.

Einstein found out how to do this, but at a cost to our intuition. He recast the equations we normally use to work out where something is or how fast it is going. These equations work very well, but they are strange. If we look at someone zipping past us in a spaceship and want to work out where they are at a given time, Einstein's equations define where that spaceship is with respect to us. But the equations tell us not just how space and distance are transformed, from the spaceship's point of view, but also time. The faster the spaceship goes, the slower we see time appear to be on the spaceship. And from the spaceship pilot's point of view, watching us zoom away at high speed, our time in turn is slower than that shown by their wristwatch. It's a phenomena known as time dilation. This is Einstein's cost to our intuition; that there is no absolute time, only relative time.

If this seems strange compared to our ordinary experience, that's because it is. These effects only make themselves apparent when things are moving very fast and very close to the speed of light (some three hundred million metres a second). Einstein's equations work very well at our slower speed of existence too, but the effects are so small for us we never notice them. You can see this by looking at the equations that tell us what position (x') and time (t') something moving at velocity v, seems to have from our point of view (at position x, and time t). The effects are magnified by a factor γ that gets bigger as velocity gets closer to the speed of light c.

$$x' = \gamma(x - vt)$$

$$t' = \gamma\left(t - \frac{vx}{c^2}\right)$$

$$\gamma = \frac{1}{\sqrt{1 - \frac{v^2}{c^2}}}$$

At everyday speeds where we move much more slowly than light, γ is about one, $\frac{vx}{c^2}$ is tiny, and times t' and t seem coincident. Positions make sense too – if you move at a certain speed for a given time, your position will be shifted by the product of those two quantities. However when velocity nears light speed, γ is large and impossible to ignore. It stretches the difference between our perceived positions and distances. It stretches the difference between our perceptions of time.

We know that the equations work very well near light speed. One piece of evidence comes in the form of cosmic rays, high energy particles that batter the Earth from outer space. When cosmic rays hit the upper atmosphere a fraction of them turn into short-lived particles called muons. These particles exist for only a millionth of a second before decaying to electrons, enough to cover only a few hundred metres at light speed. According to Einstein's world-view, however, time dilation at these high speeds means, if we watched a muon hurtling down through the atmosphere, we'd see its millionth-of-a-second life *stretched out* for long enough for it to reach the Earth's surface and be detected there (several thousand metres from the upper atmosphere). So, if Einstein is right, we would detect large numbers of cosmic ray muons at the surface, and not if he isn't. And we do detect them. Relativity works.

The story of Nowak the pilot raises what would otherwise be a simple question: how do you stay in touch when travelling? When you're travelling at near-lightspeed, communication is more problematic. Imagine that Nowak's wife beams a signal to him on the ship as it zooms away from her. Even the simplest signal, pulsed once a second, would become scrambled. Nowak would see a signal arriving less often, because he is continually moving further away, and each pulse has further to travel than the previous one and arrives a little later. Assuming that you could still decode the signal, everything would seem slower in it. The clock on the wall takes longer to tick a minute. Speech and movement slow down. Your signal, beamed back home, would be similar.

This drop in frequency causes the drop in pitch you hear when a police siren moves away from you. A similar effect causes the frequency to rise when the police speed towards you. The phenomena is called the Doppler effect in physics and affects everything from sirens to the frequency of light from distant galaxies changing as they move nearer or further away from us. It also happens when Nowak turns his ship round and contacts home. Each pulse he emits now travels a shorter distance than the one before and gets there faster than the second spacing it leaves with. The signal from his wife is similar, looking as if her life has sped up compared to his. It would be possible to correct for this if you know how fast Nowak was travelling. Typically he hasn't given her (or doesn't know) this vital piece of information, and their communication breaks down a little more.

In fairness to Nowak and the mistakes he makes (missing his daughter's birthday), considering other people's perceptions of space, distance and time can be confusing. When relativity was first proposed it wasn't possible to travel near to the speed of light to test the consequences. Theorists posited experiments in thought instead, adventures in space and time of the mind, to explore and test the equations. Nowak experiences one of the paradoxes of these thought experiments as a by-product of his work. The paradox is known as the twin paradox.

We can perform the thought experiment here, in the pages of this book. Imagine two identical twins, one of whom is an astronaut and eager to explore the galaxy. The other prefers a quieter life on Earth. The astronaut has a space mission and blasts off from Earth in a high tech spaceship travelling close to the speed of light. To the twin watching on Earth, the astronaut's time seems to have slowed down to a crawl. To the astronaut watching Earth recede into the distance, the twin seems to be living in an equally slowed-down time. Eventually the mission is complete and the astronaut turns around and heads back to Earth. The same time dilation happens – time always runs more slowly for something that moves with respect to the observer, so it seems to run equally slowly for the astronaut and the Earth twin from their different points of view. But when the astronaut touches back down on Earth, something strange has happened. The astronaut is genuinely younger than their stay-at-home sibling. If the astronaut has travelled very close to the speed of light and it was a long mission, this could be months or even years younger. How can this be? On the face of it everything seems symmetric – each twin experiences the same amount of slowing down.

The subtlety in understanding the twin paradox is in realising that Einstein's equations of Special Relativity are valid only when speed is constant. If anything slows you down, speeds you up or changes your direction, then the equations no longer apply and Einstein's theory of General Relativity is needed instead. The time dilation is the same for both astronaut and Earth twin as long as the astronaut's motion is constant. But in that instant of slowing down, turning round and speeding up again, the time experienced by the astronaut can be very different indeed. The Earth twin doesn't turn, and doesn't experience the same change in time. The difference is sufficient to explain the difference in age between twins when the mission is over – although it doesn't count any ageing of the stress induced by travel.

The twin paradox turns out not to be a paradox at all; it is a fact of relativistic life and one that even touches Nowak when he is offered a long, high speed mission by his employers. It is a lucrative opportunity and sold to him with the promise that he would experience far less than the three years of mission by the time he returned. Somewhat ruefully, Nowak reflects that ageing at a different rate to your stay-at-home partner adds one more way to drive a wedge in a relationship. Marriages don't always work with the elasticity of relativistic time.

Tether

Zoe Gilbert

IT WAS A BRIGHT night when Hark followed Gertie into the wood. His gut churned, from the vile mushroom tea Gertie had made him drink as they left the house, and from the hunch that this was a mistake.

Gertie grew wild as they crossed the hill, and the world grew wild with her. Tumbling leaves turned to waves, and the sky tilted over Hark's head, its starry eyes blinking at him. Gertie ran ahead, stretching tall as the trees. She leapt long grass that deepened into bristling fur along the hill's ridge. Hark's legs slowed, heavy as mud, but somehow, soon, the wood rushed around them, dark, dazzling.

'We're here,' Gertie called, her voice curling through the trees.

Between naked branches the stars still winked. Hark bent his head and clung to Gertie, watching the ground leaves whiten in the moonlight haze until they blazed and fluttered. There was no wind.

Gertie pulled him forward, further in, and two black trees moved to meet them. They were Guller, the bird man, and Murnon, the shepherd.

Guller held a sack, a hunch at his back. 'Glad to see you again, Gertie. You drank?' he asked.

'Hark, too,' Gertie said and went to take his arm. 'I hope he'll try tonight, I've told him what a gift it is you give.'

Guller grinned, his round eyes like marbles. He stood only as high as Gertie's shoulder, but Hark felt the urge to pull her away, to safety.

'One more to come, tonight,' Guller's voice wheedled, 'but he'll find us.' He took Gertie's hand and led them along the ash-white path, Hark stumbling behind, the shepherd Murnon a silent shadow between them.

They reached a clearing, a great bowl of moonlight, stars dewy in the dark far above. Hark's stomach bubbled like broth. He tried to call Gertie away from Guller but she was opening the sack with him, pulling out bundles that unfolded in their hands. He knew what they were, from Gertie's telling. She touched the feathers so easily, it made him shiver.

Murnon the shepherd held his shoulder. 'Steady,' he said, but the word echoed, far away. Murnon's face was old, the lines in it furrowed deeper by the dark and bright until it seemed ancient stone. In the silver light, they all four sat, waiting.

'Don't you want to see? The kites,' Gertie whispered, but Hark would not put out his hand to touch the mound of feathers in her lap. She turned it over, and spread the wings wide, so he could see the clean dried skin of the underside. He watched her take the white stone she wore on a thread around her neck and tuck it into the head of the kite skin. When she leaned and kissed Hark's cheek he felt the heat of her breath, and the brush of a wingtip on his bare arm.

'No sign of Tod Frost yet. So, you first, youngster?' Guller said, and Gertie stood, a tall, thin shape beside the little man. Guller unspooled a length of cord. He tied one end around Gertie's waist, and the other around a tree trunk at the clearing's edge, leaving the cord loose between them.

'What's that for?' Hark whispered to Murnon.

'A tether, lest she lose her way,' Murnon said. 'Brings her back.'

Hark could hear the thrill in Gertie's breath, deep gusts that prickled alarm in his head, even as Guller began to whistle. Gertie had told him of this, too, how Guller made birdsong in his throat, but a chill struck him, to hear the flute notes of a red kite flowing from a man.

The sound rose, it sent his mind high, clear of the trees,

and he heard the whistle of other kites, threading the air. Gertie spread the wings of her kite skin as wide as her arms would let her. Her knuckles shone white, gripping feathers as the kite skin rose, until she held it high above her head. Hark's stomach clenched as he watched her.

Wing shadows flooded the floor of the clearing and the kites called on and on. Gertie leaned backwards, the cord pulled taut from the tree, the wings in her hands seeming to beat as the kite strained upwards. The stars quivered. Hark saw the whites of Gertie's eyes in the moonlight. Though she held the wings high, her legs trembled. He stood, to go to her, to make Guller stop what was happening, but Guller shook his head. Soon, Hark saw Guller tug on the cord, pulling Gertie forward. She dropped the kite skin and fell. Guller caught her shoulders and eased her down. She smiled as if in sleep.

Hark bent over her. 'Gertie,' he tried, but Guller hushed him.

'She'll wake. She's been a long way, up there.'

Hark felt the chill of the wood's silver breath, the trees bending closer to stare at Gertie. Her eyes opened. 'What was it like?' he asked, but she sat up and pressed her face hot into his neck.

Guller was beginning with Murnon now, tying the cord to his waist, the kite's whistle trailing from his lips.

'Stars. Bliss,' Gertie whispered. 'You must try, Hark. It's just joy, flying up so high, the whole sky turning about you, no body to weigh you down. I wish I could stay up there forever, not just a few hours.'

'But it was moments,' Hark said.

'So will you try it? I want you to feel like that, better than anything.'

'No. Let's go home,' he said, but Gertie was looking past him, at a figure taking slow steps into the clearing. It leaned against a tree and watched, while Guller spun out the kite's song and Murnon swayed, his own kite skin stretched high.

Guller tugged the cord and caught Murnon as he fell, just as he had with Gertie. Then he beckoned the bent figure over.

'It's been a long while. But you've made up your mind?' Guller said. The man nodded.

The wide grin faded from Guller's face. For the first time, he looked solemn. The two of them clasped their hands together in greeting for a long time.

'Is that Tod Frost, the fisher?' Hark asked Gertie.

'He's not on the boats any more, just propped up in the ale room most days. He's come to the woods with Guller long before I ever did.'

'Isn't he too old for this?' Hark said. 'I'd rather be home, warm. Come on, Gertie, I've had enough.'

'Wait,' she said and turned to watch the two men. Tod Frost held the bird skin unsteady in the air, still nodding as Guller looped the cord around his waist. The eerie cries from the sky pierced Hark's head, and Guller's whistling grew higher, sweeter than before. The shaking in Frost's arms calmed, and the wings he held stretched wider. It went on, all of them held in the web of sound, wing shadows crossing the clearing. Still Guller did not tug on the cord, to pull Frost back to ground. Hark felt Gertie's hand gripping his own. He heard her gasp when Guller took a knife from his pocket and cut clean through the tether.

*

Hark's head spun, and his gut roiled, as he and Gertie stumbled home across the hill. Before they reached the village, he wretched and emptied the burning tea from his stomach.

After Guller had cut the tether, he had laid Frost down, still but warm, breathing, in a bed of dead leaves. He had placed the kite skin under his head. Frost was already gone, he had said, and it would not take long, here in the wood, for his body to turn cold. The burial would be done later.

'Up there, forever?' Gertie had asked, and Guller nodded. 'I didn't know you could do that,' she'd said, and the joy in her face had opened a well in Hark that filled with cold, dark water.

*

'Why not?' Gertie asked, the next day. She stood below Hark with the bucket while he climbed the ladder and painted whitewash onto the wall of the house. The low autumn sun warmed his back, but Gertie's mood, and her words, stirred the dark water that still lay cold inside him.

'Don't be a dunce,' he said.

'But why not? If you could choose endless joy, bliss that never, ever wore out, wouldn't you?' Gertie said, waving the bucket in her hand and sloshing whitewash into the weeds.

'I'm happy enough.'

'You wouldn't be if you'd tried it. You'd know then, what real happiness was like.'

'I do know.' Hark leaned down to dip the brush.

'What then. When do you feel it?'

'When I sit with you by the fire. When I eat a good bit of roast meat. Or just this, sun in a blue sky and making a good job of it.' He turned to dip the brush again, and wiped sweat from his forehead. 'Don't sneer, Gertie.'

'Those are such small things. And besides, you might lose them.'

'I might lose you?'

'You'd be alright without me. You always grumble that I'm a misery, anyway.'

'I'd like it if you were happy. If you could be.'

Gertie's smile was sly. 'Well, then. That's how I would be, if I went up, flying in the stars with the kites, and never had to come back.'

'And what about me?'

'You need to try it too. You'll understand. Tonight.'

'You're going back again, already?' Hark stared at her.

She put the bucket down. 'Of course,' she said. 'If you'd any sense at all, you'd come.'

'I'm the only one with any sense,' he called after her as she went into the house and slammed the door shut.

*

Hark lay awake in the darkness. Gertie had been to the woods with Guller many times before, but never twice in a row, and never with this new knowledge. He pictured the old man, Tod Frost, lying in dead leaves. He pictured Gertie the same way, and shivered in the bed. Stubborn, hot-headed Gertie, who had been his best friend, had both frightened and thrilled him all through their childhood.

He let himself admit he was frightened now. She hadn't denied that he might lose her. He'd always felt it, the shadow side of their friendship. Now, he had got himself this house, and if he asked Gertie to marry him, he thought she would agree. She was already here, after all. Even if that was only because she'd defied her bullying father and fled, leaving behind her family in their draughty mill on the marsh. She trusted him. And he loved her. And if he was careful where she was reckless, if he planned where she ran rashly in, then he believed he could do her good, now they were grown. He had never forsaken her, through years of mischief and wildness and slumps of despair.

It was dawn when he was woken by Gertie, crawling into the bed beside him. She was cold, shivery, and he rubbed warmth into her thin back.

'It was so beautiful, Hark,' she whispered. 'I was so happy.'

He held onto her, keeping as still as he could, until she slept.

*

He'd finished with the white-washing and was stacking stones to repair the half-tumbled garden wall by the time Gertie rose. It was another fine day, and Hark was pleased with his work. The house that had been a wreck when he took it on would be ready for their first winter together, just him and Gertie, cosy as doves.

'Why are you bothering with that?' Gertie asked, when she found him heaving a large stone into a gap. She looked pale, and weary.

'Same reason I bother with anything. For you, to make our life good.' He pushed the stone, to wedge it into place, and it rolled and fell on the far side of the wall.

Gertie huffed. 'You shouldn't.'

He followed her into the house where she sank into a chair and leaned her head on her arm. 'You'll get like an owl, up all night and asleep all day,' he called, while he washed his hands. He heard her laugh.

'It's a kite I want to be like,' she said, when he came and sat beside her.

Hark breathed deep. 'Please, Gertie, will you not go again? It's a worry, to me.'

'And what about me?'

'You'd be happier without that. Let me make you happy, not Guller and his foul tea and dead birds.' He thought of Tod Frost again, and took another deep breath. 'I do love you, Gertie.'

She sank lower in the chair. 'If you loved me, you'd let me do it. You'd let me choose the same as Frost, wouldn't you? You're just being selfish, wanting to keep me here for yourself.'

Hark grabbed her hands, though she tried to pull away. 'You want to be like Frost, a corpse in a heap of leaves? You don't know that his soul is flying forever, that he has this bliss you keep talking about. It's trickery. Guller is playing with you.'

'Guller helped him,' Gertie said, fierce. 'And it's not a trick. It feels real.'

'As real as this? Your hands in my hands, here, in our house?'

'Yes. Real, and beautiful. Frost is lucky.' She spat the words, and Hark wanted to twist her hands, to force her to see what was real, what was not, but he couldn't hurt her.

'This is real, too. And you are happy, sometimes,' he said.

'But I can't choose when, or even if it happens at all, can

I?' Her face began to crumple. 'Don't deny me it,' she said.

When she wept, Hark held her shoulders and soothed her, and when that night she tied on her boots and set out for the wood, he did not stop her.

*

Every night for a whole week, Gertie went to the wood, and came home at dawn to sleep until noon. Every day, Hark hunted for things to cheer her. He begged a sapling from the orchard keeper and planted it by the house, a promise of apples. He picked late blackberries and baked them as he'd seen her do. He chopped logs and kept the house cosy with fire, because Gertie liked to be warm. He tried to talk to her, and make her say that there were good things in life, so he could tell her he would always bring them. He would do whatever she wanted.

But Gertie would not go along with it. Always, she said there was only one way to be happy, that only a fool would not take it. She did not eat the food he gave her. She sat in the chair, or lay in the bed, and waited for dusk. One night, ragged with despair, Hark tried to hold her back when she got ready to leave. He stood before the door, gripped her arms to her sides, and shook her.

'Stop it, Gertie,' he shouted. 'It's a trick. I need you, here with me.'

'This is punishment,' she said, 'and I don't deserve it. Let me go,' and she scratched at his arms until he gave in. 'And you don't need me. That's a trick, too,' she said, and she ran from the house.

Hark leaned in the doorway, too ashamed to follow her, the night air cooling his cheeks. He looked at the stars, for the first time since he had been in the wood with her, and they had seemed to watch. Gertie was right, they were beautiful, but they were cold, and silent.

He tried to eat but couldn't. He dug inside the trunk for more blankets, to warm Gertie up when she came home, and found the bottle of Guller's toadstool tea. The smell of it made his stomach turn, and he remembered how strange the world had seemed, when he'd drunk from it, how the sounds dimmed or echoed, and the wood lost its familiar pattern in ripples. If he had wanted to believe in magic, he could have, then. But magic was dishonesty, and Guller was a liar, even if Gertie enjoyed the lie. It could not be real.

He took the bottle outside, ready to empty it onto the ground, but the sound of a kite's whistle reached him and turned his blood icy. His head flashed with pain at Gertie's words as she'd fled the house, and then they would not let him alone, the untruth of them. Gertie was making a dupe of herself, and he had not tried hard enough to stop it. He threw the bottle into the grass and began to walk, then run, up the hill towards the wood.

It was easy to find the clearing, though the light from the moon was not pearly as it had been before. There was a rustle in the undergrowth nearby, and he called, 'Gertie,' only for the wood to fall silent. He turned in the middle of the clearing and then he saw her, lying beneath the low branches of an elder bush. She was smiling, but when he shook her, she would not wake. There was a kite skin tucked beneath her head.

★

Gertie had still not woken by the time Hark reached the house, his arms roaring from carrying her, his heart thudding as he ran and scrambled down the hill. When he had laid her down, he wet her lips with whisky, then poured from the cup into each of their mouths. Her eyelids did not even flicker. He nested her in blankets, whispered her name, then yelled it. The beat of blood where he touched her neck was steady, her breath quiet but sure.

'What have you done?' he asked her, and then turned the question on himself, over and over, all through the night.

★

When dawn came, he hoped the light might stir Gertie, but still she slept. Though he called her, and shook her, he no longer believed that she would wake. It was to Guller's house he ran first.

'Tell me how to bring her back,' he shouted, as he pushed the door and fell into the room. It stank of decay, and feathers shifted across the floor in the draught. Guller stood at his table, cleaning a bird's skull. He looked at Hark as he spat again on a cloth and dabbed at the bone.

'You can't do that,' he said, his voice singsong like a child's.

'Then you do it,' said Hark.

Guller shook his head and smiled. 'Not me, neither. She's gone.'

'She's not dead. She's only asleep. You're a liar.' Hark stepped closer to the table, where the bones and innards of a large bird were spread in stained heaps.

'She was sure, you know. You should be glad for her,' Guller said, his round marble eyes holding Hark's gaze. 'Bliss.'

Hark reached out and dashed the bird skull from Guller's hand. It skittered across the floor in pieces.

'Careful,' Guller said.

'If she doesn't wake, if she dies now, you've killed her. Poisoned her, with that drink of toadstools, and with lies dressed up as magic.'

'But she lives.' Guller moved away from the table to peer up at the sky through the rimed window. 'It's only that she's no need for a body now. When it grows cold, it will make no difference to her. Bliss,' he repeated. 'Wouldn't you wish her that?'

'I'll tell everyone what you've done. Her father, all the village,' Hark said.

Guller opened his arms, as if in belated welcome. 'Spread the word,' he said. 'I'm the only one who can give them stars, all that joy, up there. Elders, youngsters, all kinds come.' He grinned, his eyes rolling before they fixed on Hark, sidelong. 'I can make you see why she wanted it. The price is only small,' and he jingled the coins in his pocket.

'But it's a lie, you trick them,' Hark yelled. He kicked at the table and the bird bones shuddered. 'You call it a low price. You've fooled her into giving up her life.'

'No, no,' Guller said, and he laughed. 'You should try, you'll see,' but Hark was already out of the door, spitting the stench of the house from his mouth.

★

If it's a lie, she will wake, he told himself, as he hurried back home, wincing at the sun that dazzled his tired eyes. If I keep her warm, feed her, give her drink, she'll come round. But the blackberries he mashed for Gertie only lay on her tongue, and the milk he poured trickled from her lips.

He searched the kite skin, which he'd carried home with Gertie, for clues to undo the spell. When he nudged Gertie's white stone from inside the kite's head, a slip of paper fell out with it. On it was scribbled, 'Gone to live in the sky,' beside a scratchy drawing of a tree with two thick branches. It was the one they used to climb, when they were still children, and Gertie would dare Hark to leap from its crook.

Holding the kite skin in his hands, he felt the power in the wings, the urge to fly, even though it was empty. He tried to picture what Gertie believed she saw, when she stood in the wood, her eyes filled with night sky. He could not go back to Guller. Then he remembered Murnon the shepherd, who'd hardly said a word there in the clearing.

Hark found Murnon up at the edge of the sheep field, sitting in a nook in the stone wall. Cloud hung low now, above the hill, and no birdsong disturbed the air. The shepherd nodded when he saw Hark leave the path to join him.

'Not as foolish as they seem, those goings-on with Guller,' he said, before Hark even asked his question. He shifted and made room for him to sit on the wall.

'Why do you do it?' asked Hark.

Murnon pointed behind with his thumb, at the huddle of

sheep in the field. 'I'm glad enough to do my work, but I've a hungry mind. I know my flock well. My brother, he shares the work, but he's no wish for more, like I have.'

'More?'

'Something greater than this. Joy, and joy that harms nobody to take. Guller means no ill, and folk need some relief.'

'But if it's a trick, the toadstool tea and making you believe something that's not real?'

'That's one way of seeing it. But if something feels real, and if it doesn't fade, what's the difference?'

Hark shook his head. 'It is different. It's like dreaming. Would you spend your whole life asleep, for a pleasant dream?'

'Perhaps. Life passes in a dream sometimes. You've no more rule over your fate than if you were dreaming, when you think on it.'

Hark thought about the house, all the mending and patching he'd done, all to make it fit for Gertie to share with him. He'd believed he was making his fate, that he knew how life would be. 'If you really think that, why haven't you done what Tod Frost did?'

'And your Gertie.'

Hark's throat tightened.

'I've a reason,' Murnon said, 'and that's my brother, Firwit. He'd be useless as half a horse on his own. I do the flock, see, and when it comes to it, the slaughter, the meat. Firwit handles the skins, wool, parchment. Same in the house. Left to it, he'd not know what to do. Shared it all too long. If he were to pass, before me, that'd be different, I suppose.'

'I wanted to share, too. With Gertie,' Hark said.

Murnon nodded. 'Well. She liked her freedom.'

Hark bent forward, his face pressed into his hands, his palms hot and wet. He felt Murnon grip his shoulder, as he had in the wood.

'You can scorn an old man's opinion, if you want, but I'd go up, once, with the kites. See why Gertie did. You might feel better. You might even forgive her.'

★

All the rest of that long day and night, Hark sat with Gertie, or paced the house, finding tasks that needed doing, then seeing the pointlessness of doing them. She was not going to wake up. She would drift away. He didn't want to leave her, but neither did he want to be there when the last life went from her body. The only comfort he could find, in the end, was that if she truly believed she was in her blissful starry sky, whether it was real or not, then she was not suffering with him. It was hard to keep hold of this thought. It sank over and over into the deep black well.

Finally, he spoke to Murnon, and agreed to come to the wood the next night. He found the bottle of brown tea lying in the grass and drank it down, standing before Gertie, to show her he was doing what she had wanted. Then he waited for the world to begin to ripple as it had before, tucked the kite skin under his arm and wandered over the darkening hill, under the blinking eyes of the stars.

By the time he greeted Murnon and Guller, the wood had become a whispering maze, and he followed them with halting steps between black trunks that loomed, the reaching fingers of bramble tripping his feet. The stars that crowded above the clearing were sparking, now that the moon had waned to a sliver. Hark wanted to vomit, but held his stomach, willing the tea to help work Guller's spell. He willed it to conjure Gertie, bold and joyous, before it was all over with.

He could not make out Guller's features as he tied the cord at his waist. He seemed to grin and frown, go from kind to scornful, with no pattern that made sense. Hark heard Guller's words as if from within his own head.

'You've put something of yours inside the skin?'

'My handkerchief.'

'Use that. Put yourself inside the kite. Let it take you up. I'll keep you tethered, here, till it's time to return.'

Hark did not tell Guller that his handkerchief was wrapped around Gertie's white stone, in the kite's head.

He closed his eyes and listened for the whistling. The thin quivering sound washed away all thought. There were just the feather tips in his fingers, the wing shadows wheeling, somewhere, everywhere, the call to rise. The kite skin spread wider, the secret beat of its wings thrummed through his arms, and the clearing was gone, closed into the wood below him.

It was true. He hadn't words for anything then, but afterwards he had to call it bliss. A sweetness familiar and new, a delight with no shape. How long he flew, drinking in the pleasure of starlight, he didn't know. In the last moment, as he felt himself falling, the tether pulling him down to earth, he heard Gertie's voice, close in his ear and then fading far above. *Stay*, she said.

Hark sat up, was sick into the dead leaves, and lay back down. The stars still blinked at him, but more kindly. Murnon and Guller murmured somewhere nearby. He lay, waiting for Gertie, for her voice, some trace, but there was nothing after all.

'Time to go,' Guller said, helping him to his feet. He picked up the kite skin and held it out to Hark. 'Keep it,' he said, 'For the next time.'

Hark shook his head.

⋆

After he had taken Gertie's note to the mill on the marsh, Hark carried her body back across the hill to the woods. The stars were hidden now behind wads of cloud, and the night whispers of the wood had faded to a familiar hush. He chose a place where no hunters would walk, where the beech leaves lay deep, and their red-brown glow suggested warmth to him. He laid Gertie down.

'I know now,' he said, as he tucked more leaves around her. 'I still think you're wrong, but you've always been pig-headed, so why would you listen to me.' He sat beside her, with her white stone in his hand. He waited a long time there in the greying dusk, letting go his last hope, before he set off home again.

He could see the white-washed house as he crossed the hill, half-made and empty, gleaming as the world around it faded. He dreaded the dark night in his bed, and worse, the plain light of morning. The well water sank through him, the cold of this life without Gertie. He did not know how to begin without her, yet, only that he must.

As he came closer, he saw a shadow lying across the stone step at his door. He stared down at the kite skin, wings folded beneath it, the head tucked down into the empty breast. He stepped over it and pushed into the dark house.

The grate was cold, but he could not face the brightness of a fire. Instead, he lit one candle and placed it where the light would not show the heap of Gertie's blankets on the chair. He was weary, but when he lay in the bed, he could not close his eyes. He saw the stars, high up above the rotted roof of the house, up above the cloud. They would always be there, comfort, bliss, waiting. He climbed out of bed and went down the stairs. He opened the door, picked up the kite skin from the step and carried it inside.

Afterword:

The Experience Machine

Prof. Jonathan Wolff

University College London

Robert Nozick's 'experience machine' thought experiment is designed to cast doubt on the philosophical thesis that all that matters to us is the nature of our subjective experiences. This thesis is generally associated with utilitarianism, an idea first systematically defended by Jeremy Bentham, which defines 'the good' in terms of pleasure, and further suggests that we have a moral duty to bring about as much pleasure (for ourselves and others) as possible. While in itself a very appealing moral theory, Nozick uses his thought experiment to question its foundations:

> 'Suppose there was an experience machine that would give you any experience you desired. Super-duper neuropsychologists could stimulate your brain so that you would think and feel you were writing a great novel, or making a friend, or reading an interesting book. All the time you would be floating in a tank, with electrodes attached to your brain. Should you plug into this machine for life, preprogramming your life experiences? [...] Of course, while in the tank you won't know that you're there; you'll think that it's all actually happening [...] Would you plug in?'[8]

Nozick's point is that if subjective experience is all that matters, then it should be obvious that each of us should choose to plug into the machine. But if we hesitate (other than for reasons that in some way we are sceptical about the technology) then it seems that we do value something other than subjective experience. If so, then the utilitarian position is flawed. Beyond that, however, we are challenged to try to give an account of what is missing. Nozick argues that we value authenticity: we want to actually do things, rather than simply have the subjective experience and belief that we are doing them.

In approaching a short story inspired by Nozick's example, I expected to be transplanted into a futuristic world of laboratories, mad scientists, computers, test tubes and wild hair. For all I know the wild hair is there but Zoe Gilbert has very imaginatively transformed the setting to fairy tale fantasy rather than predictable science fiction. And, like the best fairy stories, it has both charming and disturbing elements. When the story is set is hard to place: forests are timeless. Are we in the deep past or the dystopian, post-technology future where parchment is once more in demand? The location is also unspecified but the names, and the need for a warm fire, suggest somewhere in northern or central Europe.

Although, necessarily, the shift means that *Tether* does not exactly parallel Nozick's thought experiment, all the essential philosophical ingredients remain. The obvious differences start from the replacement of the machine with the idea of putting on a kite skin, and soaring into the sky, or at least believing that this is what you are doing. In Nozick's example it is suggested that the decision is time-limited, and that every few years you can reset the options if you wish. This option is replaced by two types of experience. One is short, and potentially repeating, held back by the tether. The other is permanent in which the tether is cut. Furthermore, in Nozick's thought experiment you can choose any experience you want, but in *Tether* there is only one, although we know it is blissful. Fortunately, these differences do not affect the philosophical questions.

Nozick wants us to ask: why not replace our relatively mundane lives with guaranteed valued experience (in Zoe's story, guaranteed bliss)? The characters in the story have their own answers. Hark seems to be motivated by a combination of natural caution, perhaps the idea that he has some sort of duty to remain in his allotted role, and, as suggested by Nozick, perhaps a desire for an authentic life, rather than something approaching a dream state. Murnon, the shepherd, is held back by his sense of duty to his brother, the appropriately named Firwit, who cannot cope on his own. Gertie, however, takes permanent flight, yet against a background in which real life for her holds few attractions. But it is not clear, no doubt deliberately so, whether ordinary life was always so unappealing to her. Certainly it has little to offer once she becomes aware of the possibility of spending her entire life flying as high as a kite. And, that phrase, of course, makes us wonder whether we needed Nozick's experience machine to make the point. Publishing in 1974, Nozick was well aware of drugs such as LSD and heroin that promised something like the experience machine, albeit sometimes at great cost. The experience machine tidies things up, as does the kite skin, making the experience predictable and without extraneous risk.

The wider canon of science fiction is well versed in this dilemma. Perhaps most famously, Aldous Huxley's protagonist in *Brave New World,* the innocent John Savage, claimed authenticity when he rejected the pleasure-drug soma and everything else in this cosseted new world, declaring: 'I'm claiming the right to be unhappy.'[9]

Nor is authenticity limited to pleasure-giving drugs or pleasure-giving virtual reality machines. Consider the following. Out of the blue you are suddenly head-hunted for a new job in a luxurious, modern, far-away city (let's say Dubai), offering you a better wage, lots of sun, and an all-round better standard of living. The only drawback is it means moving to a entirely modern metropolis, built quickly just a handful of years ago (and funded by petrodollars – the sheer good fortune of the region's geology). As a city, it has a reputation for being soulless, devoid of personality (compared to any town or city in the UK). Do you take the job?

Where, in the end, do our sympathies lie? Returning to Gilbert's story, it is hard to criticise anyone. Tod Frost, the first to leave the tether, has what seems a worthless, ruined, life on earth. So his choice is understandable. Murnon has achieved a 'balance' of sorts: occasional bliss coupled with taking responsibility (we might call him a 'functioning addict'). All Hark wants is a normal life with Gertie. Well, he can't have that, and knowing that Gertie values bliss above him may be one reason why he chooses not to follow Gertie's path: would they even be together if he did? Gertie is clearly in a depressed state in normal life, and Hark cannot lift her out of it. So she chooses a way out. Yet the question is posed: how much difference is there between putting on the kite skin and committing suicide? Of course some sort of personal identity is preserved, but at the same time something is left behind. In Gilbert's story it is a physical body but that can be a metaphor for a previous life that is now gone.

We should not expect a short story to resolve a philosophical puzzle, but it can make us think more deeply about it. And it reminds us that literature is often in exactly the same business as philosophy and ethics, testing and challenging precisely the same hypotheses as the thought experiments.

Notes

8. Robert Nozick, *Anarchy, State, and Utopia* (New York: Basic Books, 1974), pp.42-45.

9. Aldous Huxley, *Brave New World* (London:Vintage, 1931 repr. 2007), p212.

The Tiniest Atom

Sarah Schofield

THE BOY WHO ANSWERS the door looks so much like Ted, all Frank can do for a moment is stare at him. Frank knows that the boy is called Thomas and that he is aged seven. Tall for his age, Thomas stares back eye-to-eye with Frank.

Frank pulls himself up. 'Is your Ma in?'

A wren-featured woman comes to the door. Nancy. Her mouth twists down at the corners and beneath her eyes are lavender blue crescents. 'Yes?'

Frank looks up at her. He presses his hands into his pockets and fingers the notebook segments. 'I'm here for Ted.'

Nancy's mouth draws tight. Finally she says, 'You'd best come in, then.'

He drops his haversack by the door and stoops to unlace his boots. He places them between a pair of child's shoes and a large pair of slippers, their desiccated soles slowly contracting.

Nancy is in the kitchen, holding the kettle. 'You served with him, in Arras?'

He nods.

'Was it..?'

He clears his throat. 'Aye.'

'Right,' she says. 'Right.' She turns to the range. 'You'll have tea. You've probably come a long way. I didn't catch your name.'

Thomas scowls at him around the door. Frank winks back.

'Why are you so small?' Thomas says.

Nancy twists round, wiping her face on her apron. 'Thomas Edward!'

The boy lowers his head, watching Frank through his lashes.

'Could wrestle any man down,' Frank says. 'Your dad included.'

'No you couldn't,' he mutters.

'Get out and play, Thomas.' Nancy's voice is brittle. Thomas glowers at Frank as he passes.

'He's…'

'Aye. It's fine.'

'Will you sit?'

I drop a sandbag in place on the fire-step and climb onto it. You stoop beside me.

'Bantam division?' you ask.

I nod.

'They said you were coming.' You move over slightly. 'I'm Ted. I'm knackered.' You point your bayonet. 'Over there. Watch it.'

Artillery fire arches across the metallic sky, we buckle our heads down simultaneously.

You step down and fold yourself into a dugout a few yards down the trench and place your helmet into your lap.

I wince each time the rifles bark, but you just pull your balaclava closer around your face. You unravel your puttees, slide off your boots and pull a pair of socks from your top pocket. I press myself into the side of the trench. My gun is sticky with cold. I swallow back a wad of gritty spit. Along the fire-step, men merge into the brown sludge. We're all the same – stinking, dirty, wide eyed. There's only a handful from my battalion who made it here. Short and tough like me. We've got used to the banter and neck ache. We find ways to tuck in the spare fabric of our uniforms.

The night is quiet; time dragging in anticipation. A terrier trots through the trench mud and you swat it away. I watch you rewinding your puttees, your feet pigeon-toed together. You tie a precise hitch in the tapes. You take out a notebook and pencil stub. You glance up at the night sky.

Frank reaches into his pocket. He lays the notebook segments across the table. Tide-marked and speckled, they hold the contours of Ted's body where he had kept them pressed between his tunic and vest. Frank flips one that is smudged iron brown across the page, holds it flat against the table. It springs proud when he raises his hand. Nancy's knuckles are white on the back of the chair. She turns suddenly and leaves the room.

The water rolls to a boil, spitting onto the range top. Frank lifts the kettle off the heat and fills the teapot. He is pouring the milk when she reappears. She has a needle pursed in her lips and she sits and unspools a length of thread, snapping it with her teeth. She slides one section of the notebook towards her, resting her hands over it. Then she takes a second segment and squares them together. She threads the needle and pushes it through the paper, the thread hisses tight through the spines, drawing the sections back together. Frank watches her then he scrapes back the chair opposite.

'You were good friends with Ted?'

He stares out of the window. 'Aye.'

He pours the tea. Then he gathers the remaining notebook sections and orders them carefully. He hands them to her one by one. The stitched together pages twist away from each other, like an over-ripe flower.

Their tea goes cold.

She lays the stitched up book on the mantelpiece and places the flat iron on top of it. Finished, it is a couple of inches thick. Beside it Frank notices the empty red leather cover.

'You'll stay for supper, before you go on your way.' She sets the table. Thomas comes in and washes his hands at the sink.

'My mother's coming.' Nancy lifts a china dish from the dresser and spoons dripping into it. She raises an eyebrow at Frank.

He opens his mouth, then draws a chair over to the mirror hung above the fireplace, kneeling up on it to palm his hair flat. A sour-faced woman appears in the mirror's reflection; she stands at the front door.

'Thomas Edward.' The woman's voice is like a siren. 'I hope you've washed your hands.'

She looks at Nancy but waves a finger at Frank. 'Who's this?'

'Served with Ted. He brought us some of his things…'

'Does he have a name?'

Frank wipes his mouth on his sleeve. He steps off the chair. 'My name's…' He falters. 'They call me Ted.'

Her eyes dart from Nancy to Frank. 'Strange coincidence, isn't it?'

Nancy ducks her head, scrapes the spoon against the side of the serving dish. She turns to the boy, sitting watchfully. 'Thomas, have you washed your hands?'

Later, the older woman looks on as Frank thanks Nancy for her hospitality and Nancy thanks him for visiting. Frank stands alone in the lane outside the house. He waits for the light in the kitchen to go out then he hops back over the wall and picks his way into Ted's kitchen garden. He unrolls the blanket from his haversack and lays it onto the warm soil between the potato drills. The speckled night stretches overhead. Orion's belt. The Plough. He knows about them now. He looks at them with narrowed assessing eyes as Ted had. He squeezes his arm, feels the shrapnel shifting beneath the skin.

I wedge myself beside you into the cubbyhole. 'Room for a small one.'

'Not really.' Your elbow jabs in my side. You bend double while I lay back on my arms in your shadow.

'We look a pair,' you say.

I pull a hard square of fruitcake from my pocket. Break it in two. You raise an eyebrow at me. 'Found it…' I grin at you.

You push your notebook inside your tunic. 'Ta.'

We rest there together, the night turning. Waiting in the silence.

'I'm okay,' you whisper, 'I'm okay as long as I can see the stars… Make notes…' you pat your chest. 'Keeps me busy,' you say. 'Keeps me warm.'

We exchange glances as three shells explode one after the other not far to our left. We huddle back against spraying earth. A cry comes from somewhere down the trench. 'Stretcher bearers, this way!' Staccato shots fire.

A shell lands a few feet shy of the trench. It shakes the earth, rattling my teeth, dissolving my insides. My ears sing.

'Stand to!'

Tight jawed, hard faced, you push yourself out of our dugout. I climb up onto the sandbag on the fire-step. We stand shoulder to shoulder as the shells rip over us.

Frank lurches awake, his throat raw, the taste of blood in his mouth. He holds his breath to listen. A breeze hushes across the garden. A light flickers on in the kitchen window. He sees Nancy's pale face squinting out into the dark.

You have your notebooks – you draw and make notes. You show me them, how they are sections of the same book, the binding threads unpicked so that you can distribute the book's bulk under your uniform. 'My Nancy did it, so I could bring it.' You tell me about the lectures you go to in London; philosophy, physics, astronomy, the natural world, when you can afford it, when Nancy will let you, when you earn enough from your carpentry. You show me the diagrams in your notebook. As we share a smoke, you tell me you heard about my gallantry award. I blunder over it and change the subject. But you look at me with eyes of an equal. They start calling us little and large.

He wakes in daylight and birdsong. He stuffs the blanket into his haversack.

When he turns, Nancy is standing in the garden. 'You're still here, then.' Her eyes are hooded. She bends to pull a beetroot. It tears from the ground in her tight wrought fist, earth scatters across the vegetables.

Frank gazes at the small crater the beetroot has left. When he looks up, she has gone but a hoe is propped against the garden wall. The handle points in his direction.

He weeds around the cabbages, then the carrots and pea canes. He places his hands over the smooth intimate places where Ted's hands have been on the ash handle, and hoes in unwieldy strikes. He feels the fragments grating inside his arm. He leans on the wall as Nancy hangs sheets over the washing line. Thomas circles her with a toy plane. *Like a moon,* he thinks.

'What are you smiling at?' she asks.

Later, he sits at Ted's table for lunch. Thomas props his chin up reading the morning paper, his plate pushed to one side.

'Reads it every day,' Nancy whispers. 'In case there's any news. In case our telegram was wrong…' She looks at Frank.

Frank reaches over and presses the paper shut, points at the headline. 'You see some sights out there. A shell tears up metal like paper.'

Thomas listens, his mouth open.

'Exploded earth moves like water. Your dad… if it were quiet, he'd go and look at things, up close, see how they'd misshapen.'

'Obsessive,' Nancy nods. 'Always...' Her eyes widen. 'You know. How things are put together.'

Frank picks up his china cup and turns it in his hands.

'Did he tell you about us?' Thomas says.

He puts the cup down and looks at his plate. The beetroot seeps crimson into the wedge of bread. 'Aye.' He scratches at his bare scarred forearm. 'Aye, he did.'

Thomas stares.

Frank lays his arm on the table. 'Bits of all sorts in there. Metal, wood… Feel. Go on.'

The boy reaches out hesitantly and touches Frank's skin. He retracts his hand and slides it under the table.

Nancy gets up and takes baked apples from the range.

'I was thinking of staying, just a while,' says Frank. 'For Ted…'

Nancy puts the baked apples on the table. She scoops the largest out and puts it into his bowl. The fruit is tangy and sweet. Thomas pushes the apple flesh around his bowl until it browns.

Frank goes into Ted's workshop and breathes the cool dusty air. He touches the workbench, the machinery, saws, measuring tools, neatly labeled slabs of rosewood, oak and pear. He flips a crate, stands on it and leans on the bench top.

He takes a chisel from its setting, turns the blade between his fingers. He finds Ted's whetstone and oil and blows the dust off them. He sits on the bench with the whetstone between his knees and runs oil onto it. He circles the blade against it until the metal is silvery keen. It nicks a notch of skin when he tests it on his thumb pad.

Nancy and her mother are talking in the garden outside.

'Who is he, Nan?'

'Ted's friend.'

'What regiment?'

Nancy says something that Frank doesn't catch.

'Well, where's he sleeping?'

'He's renting a room, Ma. In the village.'

'I don't think so, my lady. I've asked round…'

'Ma!'

'What's stopping him getting back to the front?'

'His arm. He's…'

'You can't keep him like a pet. Has he no family to go to?' She sighs. 'You need to get rid. Someone'll be missing him.'

Later. The light is fading. Frank has sharpened all the tools and is oiling the pruning shears. Nancy leans in the doorway. 'Supper,' she says, sweeping her hair back. 'We're going to church in the morning. Will you come?'

Frank sits between Nancy and her mother in a pew close to the front. People bow forward as if in prayer and stare at him as the organ plays.

The vicar climbs the pulpit. Nancy's mother leans in. 'Mr James is a close friend of this family,' she hisses.

Thomas catches a yawn in his fist.

'The lesson is taken from the thirteenth chapter of First Corinthians, beginning at the first verse.' Mr James clears his throat. '"Though I speak with the tongues of men and of angels, and have not charity, I am become as sounding brass, or a tinkling cymbal."'

You beckon me into our dugout. Point into the corner at a small burner. 'Made us this, from scraps lying about. It'll disperse its smoke. Just need some fuel…' you raise an eyebrow at me.

It doesn't take me long to acquire some.

You light the little flame and we set a pan of water over it. We sit and watch it. Then you flip through one of your notebooks. You hand it over.

'I designed this. Already started it at home…'

I look at the drawing. 'A clock?'

'It's called an Orrery. Miniature mechanical solar system. I worked out a sort of scale for it. Got to be right, see.' You tap the page. 'The scale isn't consistent, or the thing would have to be huge for it all to fit. I've altered it to make the planets understandable.'

'I see.'

'Nancy calls this sort of thing my hobby. I'll probably get home to find she's sold my tools and turned my workshop into a home for stray animals.' You give a dry laugh. The stove flame flickers violently. 'I'm going to make a discovery.' You speak so quietly that at first I don't follow you.

'What like?'

You look at me steadily. 'It's a ripe old time for it; new things being discovered. Shame the war's thrown it all off kilter.'

You take out your pencil stub, whittle the tip, blow the shavings into the flame. 'You heard of Laplace's demon?'

'Religious thing, is it?'

You shake your head, smiling. My cheeks burn.

'Laplace. A scientist, philosopher. He has this theory. The theory's the demon,' you look up darkly, 'because it's a frightening idea.'

'Aye?'

You don't say anything for a moment, but stare into the flame. 'All the stars and planets, all their movements are determined. You could wind time back and forward, and if you know the speed, forces and positions of them, in the past, and now, you can work out what they'll do in the future. At any time. It runs like a clock.' You pat your chest, where your notebook is. 'Laplace said if the stars and planets work so precisely, then everything else must, too. Every tiny atom. Every being. If you could invent, or find an intellect, that could see all of the past and present, know all the forces and whatnot that will affect it, the intellect would be able to tell our futures. It stands to reason…'

you say. 'The regularity astronomy shows us in the comets and what-have-you "doubtless exists in all phenomena"... nothing's uncertain.'

I lift the boiling pan off the stove. 'Not a big one for books. Rather be doing...'

You hold the mugs steady. 'Don't you think that's a frightening thought? Years ago, they thought the planets' and stars' movements reflected the anger of the gods. But clever men had the presence to record celestial happenings, began to find patterns. Laplace says it's the same with everything else. We just haven't found a way, a formula, to calculate all the movements and forces that affect us.' You drop your voice. 'Even though the great intellect that would have to calculate it hasn't been discovered yet, everything is already decided. We don't know how to know them yet, but all our decisions are already made, decided for us.'

'Aye?'

'One day...' you swill the tea in the cups. 'Men will look back and see how stupid we were for thinking we could decide our own fate.'

'"... I understood as a child, I thought as a child,"' The vicar drones. '"But when I became a man I put away childish things. For now we see through a glass, darkly... now I know in part; but then shall I know even as I am known..."'

'Will that be what you discover? This Laplace thing?'

You shrug. 'Perhaps... how it works. I've a thought it's to do with the universe being like a huge machine, able to hold all that information. There must be a way to understand it, see what's predicted. It's a bugger being here. But I will discover something, maybe the formula, how to read it, when this is all over. I have this feeling. I'm going to make it through. You too, pal.'

I see your toes have slid in on themselves.

'After... you'll come and visit us: me, Nancy and Thomas. I'll have made my discovery and we'll sit at my table, chatting about old times while Nan makes supper. Maybe we'll tell them a few tales about now. The good times...' you pause. 'Anyway, we'll all tick along together just right. I know it, pal. We'll be two old fellas with grown sons of our own one day and they'll look just like we do now.'

'Here endeth the lesson.'

Nancy's mother glares at Frank as he rubs his eyes.

'So in conclusion friends… Our men, out there, they know not their fate. We cannot know our own. The day, the hour. So make good your thoughts and hearts. And trust that all things are held in the hands of God. Now may we all join in this wondrous hymn to nature, All Creatures of Our God and King.'

As Frank leaves, Mr James shakes his hand.

Nancy steps forward. 'This is a friend. Friend of Ted's.'

'I see. You served with Ted. We all…' He looks hollowed out. 'You'll be called back soon, no doubt.'

Nancy's mother bundles Nancy and Thomas out into the sunshine, giving the vicar a look.

'Took some shrapnel…' says Frank.

'It'll work its way out. No need to sit round waiting… And you must have family of your own… who'll want to see you.' He catches hold of Frank's arm and draws him to one side. 'You know, Nancy hasn't been well. Since…' he pauses.

Frank looks out to the brightness beyond the church porch where Nancy stares into a holly tree. Her mother is fussing with the boy's shirt.

'Perhaps it's time to go… let the mourners mourn.'

Sunday afternoon passes slow and warm. Frank goes to Ted's bookcase and runs his hands along the finely bound books. The spines glow, the leather is recently polished but when he pulls one out it sticks as if it has not been moved in some time. Frank cracks the spine and flips through the pages. The words scuttle about. He sits down, rubs his face. He squints at a diagram where the labels lean forward like fervent little ants. He rests his head against the back of the chair.

You change as soon as we enter the Arras tunnels; shrink into yourself. You've been quiet all day, taken on a queer colour. 'Safe down here,' I say. 'At least compared with up there. Like home for me. Dug enough of it…'

You don't reply as you stride on, ducking your head away from the chalk seams and echoes. I dodge along the yellow-lit tunnel to keep up. In our new quarters, we change into dry clothes. Make our bunks.

It is late. Men sleep heavy. But you've been shuffling and turning since we made for bed. I'm nearly asleep when I hear your voice. A gut groan, slowly getting louder.

'Oh God, oh God.'

I turn over, squint at you. You rock like an overturned beetle, clutching your ears.

'Can't breathe.'

'What's up, pal?' I reach over and shake you. 'Are you feeling rum?'

Your eyes are wide. You are high and sour. It turns my insides to see you like this.

'Try and sleep.'

Your breath drags wetly. 'I can't… I've got to get out.'

I pat your shoulder and you dart out a hand to grasp mine. You squeeze it against your chest, where your heart pounds as fast as a machine gun. And it is then I understand where your fear goes when we're being shelled in the trenches. You simply swallow it so deep inside you that it has to emerge somewhere else.

Someone across the way sighs. 'Put a sock in it.'

I cast around desperately. One of your notebooks is tucked into your bunk. 'Imagine… Imagine you're somewhere else.' I feel you nod. 'Think of your stars…' I kneel beside you, ease your notebook out. Flick through the pages. Peer at your jottings.

'The nebulous something… I can't read your bloody writing, pal.'

You steady a little.

I skip on some pages. 'What's this… here we are; your mate, Laplace, Ted. "We may regard the present state of the universe as the effect of its past and the cause of its future…"' I feel you loosening like a bolt. '"An intellect which at a certain moment would know all forces that set nature in motion… If it were vast enough…"' I flip the page '"…it would embrace in one single formula the movements of the greatest bodies of the universe and those of the tiniest atom."' Your breathing slows. '"… Nothing would be uncertain and the future, just like the past, would be present before its eyes."'

I wait with you, until your fingers slacken and I can slide my hand free. I lie down as quietly as I can. My heart squeezes. And I can't sleep.

In the morning we take our mess tins to the kitchen. We pull back our shoulders and stand side by side.

Nancy leans over him. His jaw is tight. His shoulders are braced back.

'Ted.' She lifts the book from his hands, slides it back onto the shelf. 'I'll look after you…'

She leads him up the stairs. He stands meekly as she pulls his shirt over his head. She guides him to her bed, and he lies there while she pulls the sheet over him. She strokes his hair back off his forehead. It is a fresh summery afternoon and he sleeps like a child.

You cope during the day although you are surly and short tempered. I don't take it personal. As we go about like moles, I point out the stuff folk have carved into the chalk walls; sweethearts, dates and blue mottos.

You lose it again at night. Your spine is tight through your shirt. Huddled against the wall beside your bunk, you scrape away at it with your knife, the chalk pattering on the floor. Men are stirring.

'Ted,' I hiss. I tap you on the shoulder. 'Ted, stop.' But you won't stop. I shake you. You're solid, like a piece of stuck ore. I grab at your hands. Your knife slices my palm and I snatch it back.

'Stop, pal,' I say. 'Look what you done.' I push my bloody palm in your face. You drop the knife onto your bunk, look at me starkly. 'I'm sorry. I'm sorry.' You hold my stinging hand. You pull me into you. You hold me so tightly I can't pull away without risking waking the others. Through a gap under your arm I can see what you've done to the chalk wall. I don't understand it, some sort of maths; a triangle, a squiggle, equals, a nought. An equation. Over and over so they all merge together.

Nancy rocks him against her. Outside the moon is large and confused. He sits up and looks at his split, bloody fingernails.

She is looking at the headboard. 'See what you've done,' Nancy chides.

Frank does not understand the markings he has scratched there, but this he understands well enough: an equation, a mysterious short hand, which means this thing, plus that thing equals the next. It is fixed and certain. There is no choice. It is already decided. Fate. Unstoppable like a stream, dammed by a child, the water has to find another way. Even when it is bombed and broken, it must add up. He must make it add up. 'It should have been me. I should be dead. Ted had something to discover. I'm fixing it.'

Nancy steers him down the stairs and out into the garden. 'You can't sleep in here. You have to go; it's night time.' He stands amidst the vegetables and watches her go back into the house.

The call echoes down the tunnels. You help me gather my pack, your nimble ease returned. Your own pack is slung ready on your back.

While my guts squeeze and turn, you look like your old self again. You jostle ahead along the corridors following the arrows to the No. 10 Exit. I trail behind. The biting chill comes first. Then a growing dawn light. You are smiling as we climb the rough cut steps, blink away the freezing sleet. We emerge like ants from a crack in the dirt. I shiver. You pull me up. You look at me. You are you again, and I am me. And we charge. Machine guns rattling. Shells arcing over us. I lift my arm, shield my face from flying debris. You drag me in your wake. You are before me, and I am your shadow. We run. Stumbling over chopped up earth.

You stop so suddenly I slam into you. You spread yourself. And keep spreading. And you are falling backwards. 'Ted, pal…' I shout. But you don't turn, you just keep falling and I try to hold you, but I am underneath you. Pinned and breathing in mud. I wriggle, try to lift you off me. I see that you are now not one, but a million pieces. And you are everywhere, splintered across the field. Splintered into me.

He spends the next day in the sun. He has found a garden sieve. He sits in the vegetable patch sifting soil through it. Each fine grain, as it falls, must fall where it falls. Because this is already decided. Already known. He thinks and thinks about how he might make Ted's discovery; the formula in the universe machine.

The day passes. After supper, Frank sits with Nancy by the range. Thomas leans over a jigsaw puzzle at the table.

'What are you doing that now for?' Nancy closes her eyes. 'Bedtime soon.'

He bristles, slides the wooden puzzle pieces around.

Frank goes over to the table. 'You should start with the corners...'

Thomas hunches his shoulders. He pieces together some yellow bits of puzzle. The sun. He groups other similarly coloured pieces. Planets, Frank sees, picking up a piece and inspecting it. The paste holding the picture to the board is flaking. Thomas puts out his hand and glowers at Frank.

After Nancy has sent Thomas to bed, Frank returns to the puzzle. Most of it is completed. The star clouds, the red, blue and green planets. He clips an edge piece into position. Nancy comes over and joins him. She leans across him and slots a piece into place. Her hair brushes his cheek.

'Looks like his father, don't you think?'

Frank reaches for another piece of puzzle.

'It gets to me, that. Sometimes.' She picks up a piece then puts it down again. 'So I think of him as a little companion instead. Like a puppy or something.'

'We had a terrier in the trenches.' Frank says. 'A ratter.'

'Ah. Ted wouldn't have liked that.' She smiles. 'When I was a girl, on holiday one year, there was this old dog that followed me around. At night when everyone had gone to sleep I went out to find him. He slept over my feet. He never barked or anything.'

Frank turns a puzzle piece in his fingers.

She looks up. 'The day we left my mother chased it away, clapping at it down the street. Cowering, it was. Looking at me... She said I'd done enough damage.'

Frank slots the bit into place.

'I'll be going to bed, then.'

He looks at her. Then he nods, goes to the door and pulls on his old worn boots. 'Night, then,' he says.

She turns off the lamp. 'Night.' She is a silhouette. Her hair a static halo around her head, she tilts her face. He steps towards her. She doesn't move, but he can see she is trembling. He reaches out. She is soft, complete.

'Ted,' she moans into the top of his head.

Her breath is hot and cold on his skin. Her fingers are on his nape, tingling over his scalp. He shudders as she touches the memory of tin hat bruises. Soft skin over hard bone. Deliciously ghastly. Balmy. Woozy. She is kissing him and her teeth press against his and he kisses her back.

'Ted,' she says. 'Ted.'

He jolts her onto the table, rolling onto his toecaps, scattering pieces of the puzzle. He forces his fingers between her clothes, seeking skin, she eases his hands out, guides him up her thigh. Jigsaw pieces fall, bouncing onto the floor. She grabs at his shirt, her fingers urgent on his buttons. He strains his body against her; some clockwork desire. An elusive urge. The harder he pushes towards it, the further away it moves. She is panting like she's hurt. *I am pushing and pushing and you, Ted, are bloodied and pulped. Your stuffing punched out, palm open like a question. Your face isn't where it should be. And if you hadn't fallen exactly where you did, it would have been me, pal. It would have been me...*

The night draught chills his damp skin. A bubble of pain strains in his chest. He sobs. They hold each other until there is silence. She lets go and looks away. She slides off the table and shakes puzzle pieces out of her skirt and she fastens her blouse. Frank buttons up and shuffles his boots. A jigsaw piece skitters across the floorboards. He bends and picks it up. He collects the bits under the table.

Nancy relights the lamp and scoops together the disjointed bits on the tabletop. She sweeps them into the box and replaces the lid. Her face is golden in the lamplight.

In the morning, when Frank goes to look, he discovers his old boots have gone and in the place where he'd left them is a pair of brown leather shoes; toes slightly together, a smudge of wet polish over the eyelets. He stares for a moment. Then he slides his feet into them and wriggles his toes against the newspaper padding their ends. His feet finish where Ted's arches rose. It feels a little like he's falling backwards. He pulls the laces tight.

He spends the day in the workshop. He searches through Ted's boxes and he finds turned wooden spheres labelled Sun, Jupiter, Saturn and smaller ones labelled with letters. He cups the sun in his palm, and runs sandpaper over its camber. He sorts through the bits of the construction inside the box.

Later, he goes into the house and lifts the notebook down from the mantelpiece and consults Ted's diagram of the Orrery while he leans against the range. He skips idly forward through the pages. Near the back of the book, Ted's handwriting slopes urgently across the page, and then he has rotated the book and written more lines over his notes vertically up the page, the letters fading as the pencil tip has worn down. The words are hard to decipher. Frank holds the book one way and then the other. '...unsure about the paradox of the demon... readily... if an alternative theory...'

Nancy carries bed linen into the kitchen. 'Potato pie for supper, Ted?'

He nods and flicks the pages back to the diagram. He returns to the workshop where he clips the planets and movements together, following Ted's design. It is easy – like a puzzle that's already been made and then broken up again.

When it is done, Frank carries the Orrery through to the kitchen, where Nancy is slicing bread. He sets it on the table beside the breadboard.

Nancy beams. 'You finished it.'

'Aye, Nan.'

Thomas leans on the table and spins the planets. Frowning, he watches them turn. The movement is smooth and clear. It rotates like a toy.

I'm getting on with it, Ted. I'm putting it all together.

There's a sharp tap at the window. Nancy's mother stands in the doorway. 'Nan.' Her mouth is tight and her eyes travel over every detail of the moment. 'A word.'

Nancy hesitates, before placing down the bread knife.

Frank pushes his toes into the wadded shoe ends. The shrapnel fizzes under his skin. 'Nan was just about to serve up supper.' His fingers slide to the wooden handle of the knife. He turns it fractionally so that the blade points towards the doorway.

'Well.' Nancy's mother glares at Frank. 'I wouldn't want to interrupt...'

'Then don't.' Frank says. *They all look surprised, Ted. They didn't expect that. But then, that's me; small, unpredictable.*

Nancy's mouth falls open but she doesn't say anything. The older woman's cheeks flame, then she yelps and turns on her heels. Nancy puts a hand to her mouth. Her eyes glitter and Frank cannot tell whether she is horrified or delighted. After a moment she hurries out of the door after her mother.

Thomas looks at Frank. There is caution, but also something twinkling and familiar. *He is looking at me like you did, Ted. When you asked about my gallantry award.*

Frank shrugs mildly. Then he spins the planets again. The boy watches, his brow creased. Eventually he points and says: 'It's too big.' He turns his sharp eyes on Frank. 'Mars. And Mercury. Compared with the size of Jupiter. It's wrong. They'd be much smaller than that. So small we wouldn't be able to see them.' He pokes the mechanism. 'And the planets don't just go round in a circle like that. I've read about it. It's more complicated...'

'It's just an idea of the real thing. So we can see it.' *He's looking warily at me.* 'If we couldn't see them clearly, we wouldn't be able to understand it. How it works, how everything moves. You see...' *And, Ted, I'm explaining it to him like you did to me. And he is nodding, thoughtful, like perhaps he sees you in me. I'm here, pal. I'm here for you. And, if Thomas helps me, we'll make your discovery, about how to use the planets, figure out the formula. Everything will continue like before. We can do it, pal. We can.*

Afterword:

Laplace's Demon

Dr. Rob Appleby
University of Manchester

Laplace's Demon is the older of the two great 'demons' of physics and can be traced back to an essay published by Pierre-Simon Laplace in 1814. The actual word 'demon' never appeared in the original work; instead Laplace talked of an 'intellect':

> 'We may regard the present state of the universe as the effect of its past and the cause of its future. An intellect which at a certain moment would know all forces that set nature in motion, and all positions of all items of which nature is composed, if this intellect were also vast enough to submit these data to analysis, it would embrace in a single formula the movements of the greatest bodies of the universe and those of the tiniest atom; for such an intellect nothing would be uncertain and the future just like the past would be present before its eyes.'[10]

His demon is concerned with the role of determinism, reversibility and free will in a universe obeying the laws of Classical Mechanics and the demon has, as discovered by Frank and Ted in 'The Tiniest Atom', some startling implications.

To understand the demon, let's go back to the laws of Classical Mechanics. These laws of nature, as formulated by luminaries like Newton, Lagrange and Hamilton, are a set of ideas and equations that let us calculate and predict the motion of physical bodies. The simplest formulation is

Newton's, in which the push or pull of one body on another is described using the idea of a force. For example, we can apply a force to a bike to push it up a hill, or to a string with a conker attached to the other end, to keep the conker from flying off. Once we figure out the forces on a particular body, by a combination of experience and insight, Newton's equations let us figure out the motion in the future if we know where and how fast it is going at the present time. If there is no force then the theory simply tells us the position and speed in the future by applying the equations. This is extremely powerful. For example, when combined with the theory of gravity by the same Mr Newton, we can put satellites into orbit and even send men to the moon.

The laws of Classical Mechanics have two very important features built into their structure. The first is reversibility, meaning time can run either forwards or backwards to show how the motion evolves. In some sense, the equations do not care about the direction of time and work either way. The second feature is related to the first and is called determinism. Essentially this means that knowing the position and speed of a body at some time is all we need to know to figure out exactly and uniquely its past and future motion. So if I know the position and speed applied to a rocket now, then I know its position and speed precisely for all time simply by applying the equations.

This apparently innocuous feature has a startling consequence. Imagine our universe is made of tiny particles that obey these laws of Classical Mechanics, and that the current state of the universe in its entirety is made up by their positions, and speeds at a given time. In other words the state of the universe is essentially a snapshot of all of these particles at some time. If they obey the laws of Classical Mechanics then it's possible, knowing the current state, to calculate the state slightly in the future and slightly in the past. Hence it's possible to figure out the positions and speeds at any time, both in the past and in the future. They can be calculated in an entirely deterministic way. This is the essence of Laplace's demon, which is an entity that can somehow know all of these positions and speeds and hence know (albeit through a fairly complicated calculation) the state of the universe at any time in the future. Pretty disturbing stuff, when we consider the notion of freedom for those people living in this universe and made up of these little particles! This is sometimes called the clockwork universe, comparing the universe to a mechanical clock. But where does it leave free will?

These startling conclusions on determinism and free will are the essence of Laplace's demon. But they only apply in a universe that obeys these Classical laws. The standard way out is to appeal to quantum mechanics, where inherent indeterminacy means it is impossible to know precisely the position and speed of a particle at a given instant of time. This uncertainty is built into the theory, which our universe seems to obey, and so it is simply impossible to know the current state of the universe. More recent arguments against this demon have come in the form of information theory and the ultimate computational power of the universe, which hinge on the amount of information it has been theoretically possible to compute given the current age of the universe. Whatever the possible resolution, there are certainly enough counter-arguments to hang onto free will for the inhabitants of this universe for at least a little while longer.

Notes

10. Pierre Simon Laplace, *A Philosophical Essay on Probabilities*, trans. by F. W. Truscott & F. L. Emory (New York: Dover Publications, 1814; repr. 1951), p4.

Red

Annie Kirby

I HAD THIS DREAM, and when I woke up the world was black and white. First thought: I was still dreaming. Laurel lying grey beside me. But warm, snoring. Not a dream then. Rainy outside, perhaps, but sunlight creeping round the curtain edges, bedroom still grey. Switched on the lamp. Useless energy-saving lightbulb. Closed my eyes, the insides of my eyelids legitimately grey.

Downstairs, Laurel making breakfast, crash clatter bang. Kitchen grey, not yellow and blue. Laurel cooking grey stuff in a pan. 'What is that?' I asked her.

Laurel laughed. 'Bacon, doh! I think you need some caffeine.' Sunk into a chair that used to be antique pine.

'Laurel, there's something wrong with my eyes.'

Optometrist, ophthalmologist. Nothing found. Brain doctors. No stroke, no tumour, no epilepsy. Mind doctors. Yeah. 'Tell me about your dream, Alice, the one you had right before you lost your colours.' I described the dream using colours I didn't remember. I knew they were the colours in the dream, but what those colours looked like, I could only imagine.

A highway, with fields of giant tulips, brightly-multi-coloured, either side. Sky-blue sky. Naked men, silvery-pale, falling from the sky, scarring the blue. Falling in slow motion, dozens of them, arms and legs outstretched in circular contraptions, glittering gold. Circular prisons, not flying machines. They fall into the tulip fields. Their faces all the same, like my father's, waxy white. Tulips sway above my head.

Push my way through a forest of stems, stench of green in my face, to a clearing with a claw-footed bathtub. Lie down in the tub. Dusk falls, bathing me in red at first, then leaching the colours.

Woke up to a black and white world.

Weeks, months, a diagnosis. *Psychogenic achromatopsia.* Put less kindly, hysterical colour-blindness. Pills to fix my mind, not my eyes. I refused. Laurel cried. I took them. Her tears had no colour. Were tears always colourless? 'You're like Mary, the neuroscientist in her black and white room,' my doctor said. I looked Mary up online.

The doctor was wrong; Mary's the opposite of me. I'm real and she was a story invented by philosophers. I lost my colours; Mary's were concealed from her. She was imprisoned in a black and white room. I'm free in a world full of greys. She was smart, studied science, learnt all there was to know about how the brain and eyes see colour, but never saw a colour herself. I have a lifetime of colours, a C in GCSE biology and an undistinguished career in graphic design. She had a destiny – to emerge from her black and white prison, see a red rose and give the philosophers something to debate. My destiny, if I have one at all, is to drown in seas of grey wondering how any of this happened. Mary and I are nothing alike.

Long term sick-leave. Laurel's extra frown lines, snaking greyly across her forehead. My house unfamiliar with its new bleached theme. I watch black and white movies and wallow for hours in the tub. The bathroom was always monochrome, so mostly unchanged, a comfort. I like to sleep, to dream, because in dreams I get my colours back. A temporary respite, but still.

One afternoon, cocooned in a steamy bath, I dream of her, of Mary. A room with no windows; not black and white. Purple walls, painted yellow bookshelf, flat screen TV in the corner. A locked door. A girl, Mary I presume, in a paisley dress, sobbing into bedsheets, curls of brownish-black hair splayed across the pillow. A pattern of violet flowers on the coverlet. She rolls over, revealing her face, eyes obscured by chunky black goggles with steamed-up shaded lenses, elastic strap disappearing into her curls. A child still, twelve, thirteen perhaps. Wipes snot from her nose with her arm. 'You're new. What have you brought me? I shrug, show her my empty hands. You've brought me a key? I'm not supposed to go outside yet, I've still got more to learn. Years of study left.' Hints of antipodean uptalk in her speech. I look down at my hands and I'm holding a key. Old fashioned, copper, a curlicue. I unlock the bedroom door.

Stairs going up, so we go up. A yard. Blue sky, white fence, green lawn. Buzzing bee, summery scent of grass. On the lawn a claw-footed bathtub gleams in the sun. In front of it, a single red rose grows, sways in the breeze. Mary lingers in the door arch, inhales, strides onto the lawn. Unclips the strap, exhales, lets the goggles fall. Face blank, imprint of the goggles on her skin. Eyes dark brown, flicker from sky to grass to fence to bathtub to rose. Me, impatient. This is where she is supposed to say 'Wow'. She goes to the rose, brushes a petal with her fingertips. 'Oh, Alice,' she says, tears falling. 'I can't see the colours. The rose is still grey. All of it is grey.' Everything she says sounds like a question.

I tell Laurel to concrete over the garden; at least it will be honestly grey. Getting skinny, no appetite for colourless food. Laurel cries a lot, says I'm a zombie. Her eyes, which used to be something, some colour I loved, are stones. I guess they are red-rimmed too, but I wouldn't know.

Mary, limpid on the bed. 'What did you bring me?' In my hands, a key. Plain, black, shiny. 'The key to my wardrobe? I already have one of those.' I slide the key into the yellow-painted wardrobe door. 'This key is special.' I take her hand, cool, push through rails of fur coats, woollen shawls, tripping over shoes and boots, inhaling dust and mothballs. 'This isn't my stuff,' says Mary. A faint, fresh chill from the back of the wardrobe, whiff of pine. 'I'm not supposed to go outside yet, I've still got more to learn.' We stumble laughing into deep, crisp snow. The pines are racing green, the sky navy. I trip over a metal claw foot sticking out of the snow. Beyond the pines, a single red rose, lit by the silver moon. 'The wavelength for the colour red,' says Mary as she kneels in the snow, 'is roughly 620 to 740 nanometres.' Slips off her goggles. I watch her face. 'Well?' Her breath, white mist falling on the rose. 'Wow,' she says. 'Wow oh wow oh wow.'

Laurel wants me to get a job. 'Something you don't need colours for. We can't carry on like this.' The greyness of the world is too much. I want to sleep and bathe and dream, visit Mary and be in a world of colour, even if the colours fade when I'm awake. I flinch when Laurel touches me with her greyness.

'What did you bring me?' Mary, sulky in her paisley dress. And what I have in my hands is a looking glass. Place it on the floor, step through it and then we're falling, falling, twisting, sods of earth tumbling, past pink worms and rusty tree roots and snoozing rabbits. We land in a mud-smeared heap in the carpark of The Wonderland Amusement Park. There's a rose, of course, peeking up through a crack in the tarmac. Mary removes her goggles. Blinks. Regards the rose. 'It's nothing new. It's exactly how I'd hypothesised red would look.' I wonder, not for the first time, why it's always red that Mary must see.

Fairground coming alive, shouts of the barkers, a Wurlitzer, thumping bass of pop music, screams and laughter, stench of diesel and fast food. Colours of the lights intensify as the sun goes down, dance across Mary's face. Mary unblinking, trying to take it all in. 'Is it too much?' She shakes her head and we run, hand in hand, through the gates. Top of the Ferris wheel, in a gondola that's also a bathtub, eating powder blue candy floss. 'How does it feel to see red? How does it feel to see yellow? Pink? Purple? Orange?' She finishes her candy floss, thinks about her answer, shrugs. Mary looks out over the twinkling colours and weeps for what she has lost. She tosses her goggles over the side of the bathtub. We watch them fall.

Laurel asks me: 'Do I want to get my colours back? Doctors ask me: Do I want to get my colours back? I do. Then think, *think*: What is it your mind doesn't want you to see?' A shadow, something dark perhaps, but like the colours from my dreams it dances on the tip of my consciousness, teasing, fades away.

Mary in her goggles, battered, scuffed, taped together. 'What did you bring me?' Mary's room, for once, black and white. A disappointment, dreaming in monochrome. Outside, a storm howls. 'You brought shoes? Are they red?' The shoes are glittery-silvery-grey. The storm intensifies; even without windows to be rattled by the wind, I sense its swelling power. Mary sits on her bed, slips on the shoes. They fit. Put my palm to the wall; a tell-tale tremor. Mary unconcerned, flexing her ankles, admiring her shoes. The rumble of the wind, like a waterfall crashing down. I shout a warning, my voice swallowed by the roar of the storm. The wind plucks the room from around us brick by brick, scoops us up; tears off Mary's goggles, the straps whipping past my head. Floorboards beneath my feet, splintering. Mary grabs my hand, pulls me onto the bed. The storm churns around us.

Silence. Sunlight. Mary's shoes, scarlet. Her bed, perched on a pile of bricks. We climb down, Mary taking care not to scuff her shoes. Her dress matches the sky. Lying in the debris of the room is a black-robed, green-skinned, wart-ridden woman, Mary's goggles embedded in her head. 'Oh dear,' says Mary. 'I seem to have killed a witch.' She yanks her goggles from the witch's head, a sucking, squelching noise, and wipes them on her skirt.

We follow a road that's very yellow, under the very blue sky, a very green city on the horizon. We come to a field of flowers, tulips, poppies, very red. Their sweet, sickly perfume narcotic. A bathtub too, cracked and discoloured, flaking paint, flower stems entwined around its feet. 'The colours here are exactly how I imagined colours would be.' Mary delivers a rare smile. 'Not how I hypothesised them, but how I imagined.' I ask her if she remembers the other times she saw colours. 'I never saw colours in my life before today,' she says, yawning. We lie down in the flowers. Sleep.

Laurel's singing in the shower, a tuneless drone that blends with the white noise of running water. I haven't heard her sing for a long time. I used to like to stand in the bathroom doorway and listen to her murdering Lynyrd Skynyrd as she washed her hair, watching the skin on her collarbone turn red from the heat of the water. It annoyed her, me letting out the steam. She said we'd get mould. I'd kiss her along that red line, suds of shower gel zinging on my tongue and she'd stop complaining about the steam. I try to remember what red looked like, that red on her collarbone, what it tasted like, how it felt against my lips but I can't. Nostalgia ripples through me. I open the bathroom door. White steam embraces me and when it clears, Laurel sees me standing in the doorway. She breaks off mid-note, sudsy body puff in her hand, water streaming off her hair and over her breasts.

'What do you want, Alice?'

She's like a slug. A glistening, grey slug against the whiteness of the tiles. It's inconceivable that she's ever contained colour, that there was once such a thing as pink, blush, nude, natural tan Laurel, her skin blossoming red in the steam. I turn away, revolted.

I sit on our bed and try to imagine Laurel in one of my dreams. I picture myself wearing Mary's goggles, waiting for Laurel's body to transform from Kansas to Oz as I take the goggles off. But my imagination fixes Laurel on a spectrum of greys, from dark ash to light ash. Ashes all the same.

She comes into the bedroom wearing a towel and deodorises her underarms, vigilantly ignoring me. I watch her get dressed. Knickers and bra, both pale. Jeans, presumably blue. A jumper, scoop-necked with metallic glints through the yarn. I bought it for her, from some designer outlet, chosen not for its label but because it was purple, Laurel's favourite colour. I try to remember what it was like to have a favourite colour, to understand a colour well enough to form a preference. Laurel combs her damp hair and slips her feet into ballet pumps.

'Alice,' she says. 'What colour are my eyes?'

I feel like I should know, from memory at least. The way I knew her jumper was purple.

'Blue. Are you going out?'

'Not blue.'

She opens a drawer, transferring neat piles of folded underwear, t-shirts and jeans to the bed. Never a girl for dresses and skirts, my Laurel.

'Brown?'

'You've lost weight,' she says, her voice thin and tight. 'You should try and eat something.'

I try to identify the colours of each item of clothing in Laurel's pile. The sunflower motif on that t-shirt must be yellow. Those jeans are maroon, I know, because Laurel's black shirt is the only thing she can wear them with, the only thing that matches. But what yellow or maroon look like, how they made me feel, remains elusive, as tantalising as a forgotten word on the tip of my tongue.

She grips me by the shoulders and put her face close to mine.

'What colour, Alice? What colour are my fucking eyes?'

Laurel never swears. She relinquishes her hold on me and stands on tiptoe to hook a sports bag down from the top of the wardrobe. She transfers the pile of clothes from the bed into the bag and closes the zipper.

'I'm sorry, Alice,' she says.

I curl up on the bed and close my eyes, so I don't have to watch her leave.

I wake up groggy, blinking away the late afternoon sunshine that fills the room. My side of the bed is rumpled, a pattern of creases imprinted on my face. Laurel's side of the bed immaculate, the way she left it, unsullied by my fitful, dreamless doze. My head aches, a twisting sensation between my temples. I decide to go for a walk.

Walking's not something I've done much of since losing colours. Being outside can be overwhelming, more greyness to press down on me. But I enjoy the warm sun as I turn randomly down residential streets and it feels good to be away from the empty spaces Laurel left behind. I pause to catch my breath and get my bearings, and there is Mary, leaning against a street sign. 'What are you doing here, Mary?'

Mary isn't wearing her usual, paisley dress but a hooded velvet cloak fastened with a satin bow. 'It's red,' she says, clutching the edge of her cloak as if she's about to curtsey. 'Red is my favourite colour.' She's carrying a wicker basket with gingham lining and a hinged lid. She gestures towards my eyes. 'Are they comfortable?' I don't understand the question. 'The goggles? Are they comfortable?' I put my hands to my eyes and discover I'm wearing Mary's goggles. Strap tight and itchy on my head, rustle of duct tape, one lens shattered, stink of witch's blood. 'Let's go to your house,' she says.

Street after street of suburban uniformity. Mary stops by the front door of a terraced house with a bay window. A sun and moon doorknocker. It's my house. Not the house I live in with Laurel, but my childhood home in Eugenia Avenue. No, I say, backing away from the door. 'No, I can't go in there.' Mary grabs my hand. 'But you must. It's the only way to get rid of the goggles.'

The door scrapes over carpet. Faint scent of damp and biscuits. Tinny music from a radio upstairs. School bag on my hip. Dirty dishes heaped in the kitchen sink; my chore, which

I didn't do. We head upstairs, dragging our heavy feet a step at a time. Bath water running, overflowing; steam clouds drifting down the hall. Radio playing 'Always On My Mind'. I hate that song. I've always hated that song. I remember, before, I had called out, 'Dad? Dad?' But this time, silence. I push open the bathroom door. The shock is the same, like a physical assault, knocking me off my feet. His face drained, lucent. I thought it would be safe. I thought Mary's goggles would protect me. Black, white and grey. Black, white and grey. The overflowing claw-footed bathtub. And red everywhere. The walls, the ceiling. Mary's fingers ice in mine. My voice. 'Why red? Why's it always red?'

Even through the goggles, I can't not see the red.

Afterword:

Mary's Room and the Knowledge Argument

Prof. Frank Jackson
Australia National University

LET'S START WITH A SHORT statement of the knowledge argument in the version based on the Mary's Room thought experiment.[11]

Mary spends the early part of her life in a black and white room. She herself is dressed in black and white. She is painted black and white all over. Her television is a black and white one (the pictures on black and white televisions have, of course, lots of greys, but we'll follow the practice of including greys when we talk of black and white). Her laptop has a black and white monitor. She never sees the colour of her eyes in the mirror. And so on. (One way to think of her situation is to imagine that she lives the early part of her life in a black and white movie of the kind familiar from late night television.) While in the room, she receives outstanding lectures on the Education Channel of her TV (in black and white, of course) in all branches of the physical sciences – physics, chemistry, neuroscience, biology etc. She is a wonderful student. She understands everything and makes all the right deductions. It seems that she knows, or can know, everything there is to know about the physical nature of our world and our minds, which are of course a part of our world. For although lectures in chemistry, for example, may go better when done in colour – it helps to use different colours for the various electrons, neutrons and protons in the diagrams explaining atomic structure – it is not plausible that it is essential to use coloured diagrams. If this contention is right (and it is a part of the argument that is rarely challenged), then she knows all there is to know about the nature of our world according to physicalists. For their view is, precisely, that the properties that appear in the physical sciences are enough to give a complete account of our world and, thereby, of our minds.

It seems, however, that she does not know all there is to know about the nature of our world. For, while in the black and white room – or black and white movie if you like that way of thinking of the example – Mary will, we may suppose, learn that people who live outside the room use the word 'yellow' when they see a lemon, and 'green' for tomatoes early in the season and 'red' later in the season (or 'jaune', 'vert' and 'rouge' if they live in France). She will see and hear them doing this on her black and white TV. What is more, given all she has learnt from those outstanding lectures, she will know which wavelength configurations emanating from the lemons and tomatoes, and impacting on their eyes cause the neural states in their brains that underpin the production of these words from their mouths, and she will know in full detail the roles that these neural states play in subjects – the connections of the neural states to bodily responses more generally, their causal histories including how we evolved is such a way as to have them, and so on and so forth. But surely, runs the argument, there is something important about the people outside the room that she knows nothing about. She is ignorant of the distinctive nature of having something looking yellow to one, the state they are in when they (or anyway those among them who are not colour-blind) use the word 'yellow' to describe a lemon. The same goes for the distinctive nature of having something looking red. And so on for all the colours. But, as we said above, Mary knows all there is to know that appears in the physicalists' account of the nature of our world and, in particular, of the people outside the room. It follows that the physicalists' account is incomplete. Physicalism is false.

That's the argument in its essentials. And there's a 'kicker'. Those who doubt the claim that despite all her physical knowledge about the people outside the room, there was something important about them Mary knew nothing of are invited to reflect on how Mary will react when she leaves the room and has the experience of seeing things as coloured for the first time. Won't she have experiences whose nature she could not have predicted in advance, and, in doing so, will realise what it was she did not know about the people outside the room until she left the room? She will realise how impoverished her conception of their lives has been. And here, of course, Annie Kirby's story is very much on point. It reinforces the thought that Mary's conception of the nature of their lives was radically incomplete. Her story makes the point by highlighting how much we would lose if we lost the ability to see colours, and how much we would regain were our ability to see colours restored. What we would lose and what we would regain are precisely that which Mary never had until she left the black and white room, and is precisely what she did know about the people outside the room while she was inside the room.

Once upon a time I was convinced by this argument against physicalism. I still think that it is an important argument against physicalism and one that cannot be brushed aside as resting on some relatively simple mistake. I also think that, setting aside its bearing on the debate over physicalism which spawned the argument in the first place, it makes clear an important issue that philosophers have to address: What's the 'before and after' difference in Mary? Of course, we know the *words* with which to answer this question. Beforehand she does not have colour experiences, afterwards she does. But what should we say about the nature of these colour experiences and, in particular, about their relation to what goes on in her brain and to the properties of reflected light?

Why did I change my mind about the knowledge argument? That's too long a story for here. The latest version,

'The knowledge argument meets representationalism about colour experience', is almost 8,000 words long.[12] But let me try and give a rough, quick sense of how physicalists might resist the argument.

A popular view about perceptual experiences among philosophers – popular these days, it took a while to take hold despite its long history – is that they are representational states.[13] When something looks thus and so to one, one is in a state that represents that the thing in question is thus and so. Nothing actually need be thus and so. Indeed, not only need there be nothing that is thus and so, it is false that something's looking thus and so is a relation to an instance of being thus and so.

Let me spell this out a bit, starting with some well-known examples of illusion and hallucination. In the Heller Illusion, two straight, vertical lines look curved, and when one tries to describe the nature of one's experience, it is very natural to talk of being acquainted with items that are in fact curved. That, however, cannot be right: there are no curved lines to be acquainted with. When a straight stick in water looks bent, there seems to be something bent in one's visual field. Again, that cannot be right: from something looking bent, it does not follow that there is something which is in fact bent, even if it is only 'in one's visual field'. Finally, when one has a yellow after-image, one seems to be sensing something yellow in front of one. Yet again, that cannot be right, read literally. There is nothing yellow to be what's sensed. There only seems to be something yellow.

Representationalism offers a diagnosis of what's happening in these kinds of cases. The diagnosis allows us to avoid the implausible positing of something that is so and so whenever there looks to be something that is so and so, while acknowledging the intuitive appeal of talking in terms of something's being so and so when there looks to be something that is so and so. Representationalism says that whenever there *looks* to be something that is so and so, what's true is that one is in a state that represents that something *is* so and so. What is it to be in a state that represents that something is so and so? Roughly, it is to be in a state that, by its very nature, urges on one the view that something is so and so. One may resist the urging but it is present all the same. Although I have introduced the representationalist position by talking about illusions and a hallucination, the position is intended to apply to all cases of perceptual experience. The distinctive feature of cases of veridical perceptual experiences is simply that they are cases where what's being urged upon by one's experiential state is the truth of the matter.

On one version of this representationalist view, the version I like, the 'feel' of a perceptual experience essentially lies in the way the experience represents things to be. The 'redness' of having something look red to one is the property the experience represents the thing to have. If anything like this view is correct, what happens to Mary when she leaves the room is not that she becomes aware of properties she was previously ignorant of. Instead, she enters into new kinds of representational states. (And what happens to the subject of Annie Kirby's story? She loses and then regains important representational capacities). The challenge for physicalists is then to give plausible accounts of these new kinds of representational states – new for Mary, not for us – appealing only to the kinds of properties that figure in their picture of what our world is like.

I am not suggesting that this is an easy challenge to meet, but if physicalists can do this, it isn't true that Mary, while in the black and white room, is ignorant of the distinctive nature of having something look yellow, for example. That distinctive nature is being in a certain kind of representational state, and she will know, given her encyclopedia-like knowledge of all things physical, all the properties of that state, while knowing, of course, that she is never in such a state herself while in the black and white room.

What about what I earlier called the kicker? When she leaves the room, won't she have experiences the nature of which she could not have predicted? Not if what has just been said is correct. Mary will have new kinds of representational states of course – that is, she herself will now be one of those who are in the representational states in question, states of representing that objects have one or another colour – but the nature of the states will be just as she expected, for their nature is given by the way they represent things to be, and she knew about that in advance. The difference is that she herself will now be one of those doing the representing. However, I should emphasise that when I say that their nature will be just as she expected, I do not mean that she could beforehand have formed a mental image of being in the state or states in question or imagined what it would be like. She will lack those abilities. But that's not a sign of ignorance about the properties to be found instantiated in our world, and so is no threat to physicalism. It's a lack in abilities.[14]

Notes

11. Here I draw on Frank Jackson, 'Epiphenomenal qualia', reprinted in *There's Something About Mary*, ed. Peter Ludlow, Yujin Nagaswa and Daniel Stoljar (MIT Press, 2004), pp.39-50; and Frank Jackson, 'What Mary didn't know' reprinted in *There's Something About Mary*, pp. 51-56. *There's Something About Mary* is a very useful collection of essays about the argument and related issues. It has an excellent introduction by Daniel Stoljar and Yujin Nagasawa that critically surveys what's covered in the volume and gives some history about the knowledge argument (which goes back to 1925 at least).

12. To appear in *The Knowledge Argument,* ed. Sam Coleman, Cambridge Classic Arguments Series, Cambridge University Press. See also Frank Jackson, 'Postscript on qualia', reprinted in *There's Something About Mary*, pp.417-420, and Frank Jackson, 'Mind and illusion', reprinted in *There's Something About Mary*, pp.421-442. The focus in 'Postscript on qualia' is on *why* one should resist the knowledge argument; the focus in 'Mind and

illusion' and in 'The knowledge argument meets representationalism about colour experience' is on *where* the argument goes wrong.

13. For an extended defence of a view of this kind, see, e. g., Michael Tye, *Consciousness, Color, and Content*, (MIT Press, 2000). There is a short presentation and defence in Jackson, 'Mind and illusion' and 'The knowledge argument meets representationalism about colour experience'.

14. See David Lewis, 'What experience teaches', reprinted in *There's Something About Mary*, pp.77-103.

XOR

Andy Hedgecock

INPUT A

Neil and Gayle are having their final argument on a warm spring evening, two days after the 1983 General Election. It's eight o'clock, the light is fading and the park is almost empty. From the bench beneath the statue of Lord Stanley, 1799 to 1869, Gayle watches Neil stride down the stone steps and across the neatly mown grass of the terrace below. 'You're acting like a toddler,' she calls after him, as he heads towards the river. He shouts something in reply, without turning, without missing a stride, but she can't make out the words.

She feels beneath the bench for one of her high-top sneakers, and then crosses the gravel with fastidious steps to retrieve its twin from a wooden slatted litter bin. By the time she looks round, Neil has disappeared. The only figure in sight is a woman standing motionless on the path below, in front of the circular fountain to the left of where Neil has walked. Gayle peers at her. She wears a long, wrap coat, possibly blue – colours are fading to dusty pastels in the muted light, so it's difficult to tell. *The knee-length boots are a bit over-the-top for June*, thinks Gayle. The woman waves. Gayle begins to raise her hand, quite automatically, and then stops herself. *Who is she,* she wonders. *Am I supposed to know her?*

Turning towards the river, the woman starts to walk away, and instantly vanishes. She doesn't reach the fountain, she doesn't leave the path and she doesn't walk behind a bush. She is simply no longer there.

Gayle stares for a moment, tugs on her sneakers, then sets off to find Neil with her laces flapping. She heads wide of the fountain, towards the East Grotto and cuts down towards the river. There is no sign of the woman, and no sign of Neil. She hurries under the old rail bridge, past the Japanese rock garden. No Neil. She pauses for breath at the Boer War memorial, before jogging to the western edge of a grass amphitheatre. At the other side is a crumbling belvedere tattooed with graffiti: through one arch she sees an anarchist's circled A, but no sign of Neil. She is about to cross the grass when she hears a raucous shout – a couple of lads in their teens are heading towards her, flourishing cans and bellowing inchoate greetings. *Probably harmless*, she thinks, *but it's getting dark, don't run, walk briskly*. She heads up the path towards the exit, her legs feeling shaky and her ears ringing. *Stay calm,* she tells herself.

As she walks back to the car park the street lights, tail-lights and headlights bleed into each other. She realises she is crying and hears her own shuddering breath above the traffic drone. Sinking into the front seat of her battered Peugeot, she slaps the steering wheel with the heel of her hand. She starts the engine. As she pushes a cassette into the tape player, Marc Almond sings: 'It was a kind of so-so love, and I'm gonna make sure it doesn't happen again.'

Gayle drives away unaware she has seen Neil for the last time. They are still avoiding each other when, two days later, he is hit and killed by a transit van while cycling to university to pick up the mark for his essay on visual perception.

OUTPUT A

Gayle sits in a conference chair in Joanne's office. Her much younger boss performs the usual ritual with her foot, freeing her heel and letting her shoe swing on her toe. This is what happens when she comes to the 'talking turkey' phase of a supervision meeting. It's the bit where consultation segues into bollocking.

'Gayle, I think the team are taking the piss and letting you take the strain. Look at their performance over the past few weeks.' She starts counting off their misdemeanours on her fingers: 'The test version of the scoring module was a mess. Kerry's coding was late and I'm not sure what Richard actually does these days. From what I've seen, his purpose in life is postponing client meetings, buying equipment of dubious value, and filing weekly expense claims.'

Gayle's throat tightens as she fixates on Joanne's swinging shoe. Her Development Director's concern is well intentioned, but that somehow makes things worse.

'It's not easy being a new manager,' Joanne leans towards Gayle and touches her arm. 'In coding terms you've come to an XOR Gate: you can be their friend or you can be their manager, not both. But don't worry; I'll help you get the boundaries in place.'

As she trudges along the beech-floored corridor to the Web Application section, Gayle wonders if the game is up. She's been at Westerman Psychometrics since returning to her Midlands hometown 15 years ago, and she's starting to flounder. It's been a struggle to get a management post, and if it doesn't work out where else can she go at the age of fifty-one?

She walks through the office door and sees Kerry's screen flip from a travel website to a Word document. Nick slams his desk drawer shut. 'Like your hair,' he says hurriedly. 'The highlights suit you.'

Gayle ignores him and flops behind her twin-screened console. 'Richard around?'

'Working next door,' mumbles Kerry between bites of an apple. 'He says he's going to pop in and surprise us with something.'

'I used to like surprises,' says Gayle. No response. Nick develops an intense interest in his screen and, Kerry's fingers begin to clatter the keyboard.

Gayle opens her mail and reads a message from Rob, her ex. He wants to sell the house and split the proceeds. 'I don't want to stress you out,' he says, 'but I need to know you're not dragging your feet'. She clicks 'Reply', and types 'You can't help it can you? Bullying is your default setting.' She re-reads her comment, types 'Arsehole' and clicks 'Send'. She deletes a string of emails selling courses, software and exhibition space, before noticing an email from her solicitor saying, 'On no account should you communicate directly with Rob.' And there's also a message from Mel:

> From: melanie.aickman@psy.ox.ac.uk
> Subject: Catching up
> Having a good week you old slapper? I'm at a conference up there in the land where the woolly mammoths roam in a couple of weeks – any chance of bed and breakfast? Lots to catch up on, look forward to boring you face-to-face … Mel

Gayle smiles. Mel is her last link with a time of optimism and possibility. It was Mel – and Neil – who gave her a bit of confidence and a sense of adventure. She can imagine how Neil would have reacted to Joanne's attempt at a motivational chat. He'd have laughed in her face and asked if lacking a soul had been an accident of birth or a career strategy.

She thinks about Neil from time-to-time – usually while trading drunken and conflicting reminiscences with Mel. Her sense of connection with her undergraduate self – the socialist worldview, the passion for Dada and The Clash, the faith that Artificial Intelligence was less than a winter away – has evaporated over the years, but the memory of her time with Neil is vivid. If she concentrates, she can almost see his face.

The door flies open and a Cyberman from the David Tennant era of *Dr Who* stomps into the room, arms stiff at its sides.

'Morning boss,' Richard's voice rasps through the helmet's voice modifier, 'this is what I'm wearing for the learning technology exhibition.'

There's a simmering pause as Kerry and Nick gawk at the Cyberman's silver boots, faux-metallic shin guards, chaps, breastplate and helmet, with its neat, geometric features.

'How much did that cost Richard?' Gayle's voice remains controlled. 'More than the hire car you charged to us last month? More than the first class rail tickets?'

'Four hundred. Cheaper to buy than hire and it'll pay for itself,' he replies in a humming, electronic monotone.

'Will it really?' Gayle walks away from Richard and yanks her scarf and wrap coat from the coat stand. 'I'm off to get some lunch. I'll be an hour at least. When I get back I expect this,' she waves her hand at Richard's silver plastic body armour, 'to have been returned, or paid for with your own money.' She barges past Richard and leaves the team in sullen silence.

She walks out of the building, over a pelican crossing and up well-worn steps to a narrow stone path between a renovated Victorian terrace and a Gothic church. As the winter sun casts shadows of the churchyard's iron railings onto the Portland stone of the houses, Gayle glimpses a blue plaque: *Home and laboratory of Dr Carl Eckhardt* – but her attention is suddenly caught by something in the air at the far end of the pathway: it seems to be full of suspended opaque particles, drifting in all directions, back and forth, but confined to the exit of the walkway. *It can't be gnats at this time of year*, she decides. As she approaches the exit, curiosity overpowers caution. She holds her left arm out in front of her face and steps from stone pathway to concrete pavement.

INPUT B

Suddenly it's warm – so warm she has to slip off her coat – and Gayle is in a city 80 miles to the North West, outside a railway station. She recognises her location instantly – the place has hardly changed at all. The station clock reads four fifteen. She walks up the driveway onto one of the city's two high streets and sees the independent bookshop. It ought not to be there, having been demolished to make way for a new mall more than five years ago. She remembers reading about the protests. She wanders in. A table of recent publications showcases *The Color Purple*, *The House of the Spirits*, *Schindler's Ark* and *Deadeye Dick*. Gayle drifts out of the bookshop and is about to pass a newsagent's when she sees a placard advertising *The Sun*. 'The Great Maggie Massacre,' it shrieks. On the shelf inside the door is a spread of front pages: 'Landslide', '144 Majority for Tories', 'Thatcher Hails a Massive Majority'. Suddenly feeling hot and dizzy, Gayle walks back into the street. She checks the time against a clock over a jeweller's shop and strolls past half-remembered shopfronts rendered exotic by their drab simplicity and vague familiarity, until she comes to a store crammed with racks of new and second-hand LPs. She's surprised by how little space there is for customers to navigate past each other in the aisles. She flips through the racks, the polythene protective covers emitting the distinctive tang, faintly like petrol, that she hasn't smelled in years.

A couple of hours later, after the shops have closed, she walks into the park in which she spent so many hours as a student. She walks down to the river, passing dog walkers enjoying the pale late afternoon sunshine. Later, as the light begins to dip, she finds herself near the fountain with four baroque figures representing the classical elements. He's there already: she can hear the crisp echo of his voice across the terrace on this clear and still spring evening. And she can see him with that self-righteous girl beneath the Lord Stanley Statue. Behind the statue is the lowering gothic edifice of the Park Hotel.

★

The girl with the shock of scarlet hair, buzz-cut at the sides, slumps onto the bench and unlaces her sneakers. She kicks one off and leaves the other dangling from her toes. Neil leans down, grabs the shoe from her foot, and tosses it towards a litter bin halfway down the gravel path. As it hits the rim and drops into the empty wooden cylinder, Neil punches the air triumphantly and acknowledges the applause of an imagined audience.

'Childish,' she laughs. 'Now get it back, I bet that bin stinks.'

'In a minute,' Neil flops back onto the bench. 'You can't escape without your shoe, and we need to talk.'

The girl clambers on top of him and, with her legs astride his, leans down as if to kiss him, then snatches her face away at the last moment. 'So what do you want to talk about?' she asks, looking down at him with a smirk and expecting him to pull her back towards him.

Neil frowns. 'What are we going to do after graduation?'

'I told you, I've got the postgrad offers from Edinburgh and Brighton and a couple of interviews. It all depends how I did in my finals.'

'I don't suppose I'll see much of you once you move away.'

'Look Neil, I wish... I wish you didn't have another year to go,' she takes his face in her hands. 'But we'll survive, you know. You're just feeling miserable because Maggie Hatchet's back in Number Ten.'

He stares at her: 'Shall I tell you what *I* wish?'

★

Gayle is hiding behind the fountain to stay out of the couple's sightline. As the quarrel unfolds she stares at an ornately carved woman whose stone feet have been eroded by decades of cascading water. She hears her younger self call, 'You're acting like a toddler', as Neil walks away from her, across the grass. Neil's mumbled response drowns in the gentle trickle of the fountain. She makes out the words 'Jealousy about mean fashion'. *But neither of them cared about clothes.* Neil walks past the fountain without seeing her and veers left into the gazebo. As he does so, she walks from her hiding place towards Lord Stanley's terrace.

★

Even though she was expecting Neil to walk away from the girl with scarlet hair, Gayle is shaken by the spectacle. And the shock prompts memories. She had just retrieved her shoes, including the one from the bin. Converse All Star, with a neat black line around the sole. Her pride and joy – she had splashed out on them, along with a couple of LPs, when Neil had disappeared for several days after an earlier row. And she recalls having to borrow a few quid for the rent from Mel that month. Now Neil had stormed off again, but then that woman had appeared from behind the fountain, and started walking across the grass towards her. She remembers how her hands had shaken as she tugged on her sneakers and surreptitiously glanced towards the advancing stranger, an older woman whose winter coat and boots looked ridiculous in that weather. She stopped at the base of the steps, ten feet below her. Then suddenly a new memory. The woman shouting: 'He's over there, behind the trees. He's in the gazebo.'

OUTPUT B

Over lunch, Neil and Gayle have been discussing plans for their twentieth wedding anniversary break in Vienna. The coffee hasn't arrived and Neil is anxious about getting back to work on time. 'I'm going to have to bolt,' he says, 'and you're going to have to drink two macchiatos.'

'The upside of being lumbered with morning lectures this term is I don't have to face cognitive science Year One after lunch,' says Gayle. 'Getting anything worthwhile out of that lot is like stirring cement. They remind me of you at that age.'

Neil grins. 'Well, some of us have proper jobs and a proper chance of redundancy, so I'll make a move.'

'I'm feeling so sorry for you I might have to cheer myself up with a visit to The Contemporary. There's a new show on: the early '80s. Our era.'

'Thatcher's era,' he corrects her. 'If you go via the alley near St Mary's, have a gander at the Eckhardt house, it's acquired a blue plaque. There's a consortium trying to buy the place and open it as a museum, but the family are dragging their heels, apparently.'

Half an hour later Gayle leaves the restaurant and takes the short cut to the gallery, along the narrow stone path between the renovated Victorian terrace and the Gothic church. As she walks in the winter sunshine, between black iron railings on her left, and Portland stone on her right, she sees the blue plaque Neil mentioned: *Home and laboratory of Dr Carl Eckhardt – philosopher and pioneer in high energy physics.*

A faint movement at the edge of her vision distracts her from the plaque. There is a disturbance in the air ahead of her, a flickering moiré pattern of floating specks. *It can't be gnats at this time of year,* she decides, and continues to make for the end of the pathway.

INPUT A′

As the couple walk towards the bench every word they say falls into place like the dialogue of a favourite film. But Gayle struggles to recall the detail of events until they actually occur. She fixates on a weathered stone figure at the base of the fountain, straining to remember the words that sparked the row, but failing. For a moment Gayle wonders why she is waiting in the wings: why shouldn't she walk past them or even sit on the adjacent bench? She knows that would be stupid, but her overwhelming sense of familiarity with the events of that night troubles her. She thinks, *If I'm by the fountain, and she's in front of the statue, where's the other one? Where's the woman I saw? Was she really in this park, on this evening, watching this couple?*

All she knows for certain is that the overdressed woman plays a decisive role in her life. *And here I am, wearing a coat and boots in May.* Suddenly she knows what she has to do.

She peeks past the plinth and sees them: Neil and her younger self sit beneath the Lord Stanley Statue. The younger Gayle begins to unlace her sneakers…

*

'Shall l tell you what *I* wish?'

'What?' her voice falters.

'I wish you weren't so eager to suck up to the suits. You'll throw everything else away for a so-called good job – you'll throw me away. You used to have ideas of your own. I suppose we're coming to a dead end.'

She rolls off him and sits on the bench. 'I know you're stressed Neil, but that's really unfair. I don't want to end up on the dole, I have to make a living somehow.'

He glances at her: 'I think Melanie brings out the worst in you. In a few months you'll be a pair of big-haired, careerist bitches with fat Filofaxes, shoulder pads and no sense of morality.'

'Don't be a tosser, Neil.' Gayle turns away from him and stares at the river. 'Since we're being honest, *my* wish is that you'd show a bit of enthusiasm for something. Your course, your friends, me? Even your politics are a pose. You call me immoral... Remind me who marched on the Falklands rally last year, while her boyfriend was at a football match. You're drifting, you're passionless...'

*

The middle-aged woman behind the fountain can tell the quarrel is coming to the boil so she abandons her hiding place and heads towards them. Neil has already reached the bottom of the steps and is setting out across the grass. She hears her younger self shout, 'You're acting like a toddler'.

Neil laughs and, as he walks away he mumbles: 'Ask Melanie about me and passion, see what she thinks.'

The woman with the blonde highlights and wrap overcoat freezes until her younger self looks down from the terrace. Then, in the silence between them, the most she can bring herself to do is wave.

OUTPUT A′

In Joanne's office, Gayle watches as her boss frees the heel of her right shoe and lets it swing it from her toe.

Afterword:

The Grandfather Paradox

Prof. Stewart Boogert

Royal Holloway, University of London

THE CONCEPT OF COMMUNICATING with the past has existed for as long as space and time have been amenable to mathematical description. This started with Einstein's theory of Special Relativity in 1905, which radically expanded our understanding of distance in space and time. Time and space, from this point onwards, could no longer be considered separately but as unified 'space-time'. Special Relativity mathematically formulated the distance between two space-time points. Two points in space-time can be causally connected, Einstein admitted, if a particle travelling at the speed of light, or slower, could travel between them. Particles travelling faster than the speed of light, known as Tachyons, under certain circumstances could allow information to travel backwards in time. So connecting two points in space-time with a 'Tachyonic anti-telephone'[15] would allow communication with the past or the future.

Tachyons are just a theoretical concept, it has to be remembered. Einstein, in partnership with Arnold Sommerfeld in 1910, explored the possibility of a particle being sent at a speed faster than the speed of light, from point A to point B, with the arrival of this particle causing a change in B.[16] Seen from the point of view of a reference frame, travelling at a speed v, this event looks different according to the value of v. At some values, the message appears to arrive at B before it leaves A; causation runs backwards, et voila! We have a tachyonic anti-telephone! Interestingly, Einstein felt that this result contained no logical contradiction in itself; but rather 'contradicted the totality of our experience' so the impossibility of tachyons was taken as read.[17]

So far, all particles predicted by the 'Standard Model' of particle physics have been observed to travel slower than the speed of light. Particles with mass – electrons, protons, neutrons, etc. – cannot even reach the speed of light, let alone exceed it. If physicists accelerate particles with mass in an accelerator, the energy of the particle continues to increase proportionally but speed does not, instead the speed gets closer and closer to a finite limit (the speed of light, c), without ever quite reaching it. Where does the increased energy go? Well, the particle actually gets heavier. As a massive particle approaches the speed of light its mass approaches infinity! There are many different types of particle within the Standard Model of physics, including some famously elusive ones, like neutrinos which rarely interact with normal matter. In September 2011, the OPERA experiment published experimental results, apparently showing that the tau neutrino travelled faster than the speed of light. It made international headlines worldwide. Physicists were torn. Some refused to believe it, like Professor Jim al-Khalili who announced, 'If the CERN experiment proves to be correct [...] I will eat my boxer shorts on live TV!'[18] Others started hypothesising about neutrinos leaving the four-dimentional space-time and taking a short cut through 'the bulk' (the other seven or so dimensions hypothesised by String Theory); suddenly the tachyonic anti-telephone, and a whole host of other time travel paradoxes, became real possibilities. By June 2012, however, careful checking by scientists at CERN and the OPERA experiment revealed a faulty component of the timing system had led to the faster-than-light results.[19] All paradoxes were mercifully avoided.

But why would it have been such a disaster if this result turned out to be correct? Let's consider determinism, for a moment. Einstein was the greatest of determinists; he liked the idea of one thing causing another, preferably through the transmission of a particle. Andy Hedgecock's story picks up on this love of determinism with its very title. Nothing is more deterministic, and causal, than a logic gate. For the uninitiated, logic gates have either one or two inputs, and a single output. Given knowledge of the inputs to the logic gate, the output can be determined. Logic gates are the fundamental building blocks for all of modern computers and the programs written for them, and they wouldn't work if the physics they're built on wasn't equally deterministic. Given a known input, physics can predict a future event, or at least the *probability* of a future event (this latter modification has been brought in by Quantum Mechanics). This causal relationship between cause (input) and (probability of) effect (output) is a cornerstone of all physical sciences. How unnerved would you be if you hit the same keys on a computer and the computer responded in a different way every time? Physics students (and professors) would be even more confused than they usually are if a calculation yielded a different result every time.

There's almost no end to the havoc this would reap. One of the central problems for time travel is even the smallest disturbance, however trivial and seemingly inconsequential, could have huge, unforeseen consequences. Nor do you need to send a 'thing' or a person back in time – like a murderous robot with an Austrian accent – merely sending information back in time – like a sports almanac – can cause problems. Consider it from the point of view of the receiver of this future information (rather than the time traveller's). Knowing that in the future your life may take some course that you don't want it to can lead to not one but three different paradoxes:

(i) You alter your life choices to avoid this chain of events, this changes the future to a more favourable one, but then you never get to receive negative info from this bad future in the first place, so you never make the change, and we go round and round, in a loop, or a paradox.

(ii) You act to change the future, but this prevents you, in some way, from ever being able to send the message back. The famous Grandfather Paradox is an example of this: perhaps the person who was going to send this future back was your grandson, and your change of course brings you inadvertently to a sticky end before you're able to procreate! In truth, this paradox is probably a subset of the former one.

(iii) You change your life's course but this still circuitously leads back round to the same result; the unwished for future you were trying to avoid all along.

Popular time travel narratives accommodate paradoxes (i) and (ii) quite easily by simply saying that at any one instant in history the universe can have two states: Timeline 1 (the original timeline, that sent the first signal back) and Timeline 2 (the universe that branched off at the point of time-traveller's meddling). Narratives are used to weaving together different plots and subplots, so this is bread and butter to story-tellers. But why stop at two time-lines? Andy Hedgecock's story takes it to the next level, with three, and neatly loops each one back, in a closed, double-looped circuit. But why stop there? Why not have an infinity of different time-lines?

Here is the problem. Physics, as it is currently formulated, requires a single state of the universe. What of quantum mechanics you ask? The classic example of Schrödinger's cat (see pp.139-158) suggests that the universe is in a superposition of different states (the 'uncollapsed wavefunction') until a measurement is made, the lid of the box is lifted, and the wavefunction collapses. Aren't these different states the same as multiple universes, you ask? And what about other branches of cosmology and theoretical physics: String Theory, Branes, Holography and M-theory? Don't they all throw up multiple universes? They do, but they are all built on the same theoretical basis that there is some causal relationship between past events and (the probability of) future ones. Signally the past (or receiving information from the future) violates this, and opens up all the paradoxes listed above.

But, you say, what if tachyons could be designed to only transfer information that didn't cause long term alterations (as in paradox ii)? Andy's story could be classed as one of these stories, as, after the second circuit of the loop, his characters are back where they started. As a scientist, I'd have to say these would be some pretty hypothetical particles. From outside of the loop, at least, they would be intrinsically unmeasurable; thus the hypothesis is untestable, and it departs completely from the realm of science and sets sail for fiction only.

But if it *were* possible? Well this would be where physics, psychology and the question of free will would all crash into one another. Getting a phone call from the future would fundamentally change everything from the moment the receiver is picked up. Imagine finding yourself in a room containing a doorway and nothing else but a tachyonic anti-television showing you five minutes into the future. On this TV you see yourself walking out of the room. At this moment, what happens to your free will? Are you compelled to walk through the door? Do the laws of physics compel you to walk through the door? What if you don't walk through the door at all?

Notes

15. Gregory Benford, D. L. Book, W. A. Newcomb (1970). 'The Tachyonic Antitelephone'. *Physical Review D* 2: 263–265.

16. This paradox, sometimes called Tolman's Paradox, was also explored by Richard C. Tolman's 'Velocities greater than that of light' in *The Theory of the Relativity of Motion*. University of California Press (1917).

17. *Einstein, Albert (1990)*. 'On the relativity principle and the conclusions drawn from it'. In Stachel, John; Cassidy, David C; Renn, Jürgen; et al. *The Collected Papers of Albert Einstein, Volume 2: The Swiss Years: Writings, 1900-1909*. Princeton: Princeton University Press. p. 252.ISBN 9780691085265. Retrieved 2 Aug 2015.

18. Alok Jha & Ian Sample, *The Guardian*, 23 Sep 2011. https://www.theguardian.com/science/2011/sep/23/physicists-speed-light-violated

19. https://www.theguardian.com/science/2012/jun/08/neutrino-researchers-einstein-right

Bright Boy

Marie Louise Cookson

HER LAST NOTE MAY have said something like this: *Goodbye, Anderson. The meatloaf should be good for a couple more days yet.*

Or simply:

You will remember to feed Copenhagen, won't you. No question mark after the 'you.' Meredith Henry did not like question marks.

Or even:

Our marriage has reached a state of maximal entropy. And you never did make the maple syrup like you promised. Yours once but not anymore, Mer.

OK, it is highly improbable Mrs. Henry would have written about maximal entropy in her goodbye note. She didn't really ever talk like that, at least not when I was around. To me, she would post cryptic, pass-agg notes like: *Good day, Terri. I enjoy the sporty look on you but wearing a belt could really help to flatter your boyish figure. Could you please ask your son to watch his speed. Thanks.* My thirteen year old, Ryder, is something of a speed demon and once careered straight into Meredith outside her home one afternoon, or outside 'Entropy Central' which Ry later took to calling the Henry's house.

'What on earth do you mean by that?' I asked him.

'Just take a look over the fence,' he said.

It seemed to be the boiling part of the sap-to-syrup procedure which was causing Anderson the most trouble. My son told him that you have to heat it up outside or you'll just end up with a sweat lodge for a kitchen. Almost every single day, even on New Year's Day, he was out there in his back garden attempting to build a fire. Ryder and I asked him if he wanted any help with this but he just replied, 'No thank you, folks. I got it.' Mr. Henry hadn't got it at all, and couldn't even stay awake long enough to stop the liquid from burning. That was the first time we saw him really lose his temper, too. It was as if all the life was gradually being siphoned out of him. I did not cry though, absolutely not. I quit crying long ago. And I swore that if he hadn't got it together by Valentine's Day, I would just have to wish for some kind of devilish little leprechaun to come along and sort things out.

'Welcome to the neighbourhood, Ma'am,' Anderson called out in a voice that sounded surprisingly young for 74. This was the first thing he said to me when Ryder and I moved to this area, almost seven years ago. 'I'll bring you and the Little Newt a batch of New Hampshire's finest Henry syrup next spring.' Anderson coined the nickname 'Little Newt' for my son that day because he was small, wriggly and curiously cool-blooded. The shape of him may have shifted since then of course but Ryder has retained that chilly, amphibious-like touch.

Just as Mr. Henry said this about Ry, I caught my first glimpse of Anderson's wife when she opened the front door and stepped onto their porch. She was wearing a tangerine-coloured, wide brimmed hat and the angle at which she had placed it on her head meant that I could only see her right eye. This eye seemed to be watching Ryder and me intently. She did not come over to say hello but instead stood for a moment longer before going back inside. How strange, I thought, so un-neighbourly compared to her husband. 'Thank you,' I said to Mr. Henry. 'We'll look forward to it.'

'You won't be disappointed. Is that a British accent I hear?'

I smiled and nodded. Mum and I swapped old Manchester for new Manchester when I was a teenager, and, well, Ryder was born just a few years later. 'Afraid so.'

'Nothing to be afraid of here, Ma'am,' he said, winking at Ryder. 'Now, let me ask you something, Little Newt.' Anderson eased himself down to my son's height with such supernatural grace that I wondered whether he might have been a ballet dancer in his earlier days, rather than, as I learned later, a post office clerk. 'Do you know why –?'

'– Do I know why they're called silver maples?' Ryder interrupted, mimicking my English accent.

Mr. Henry looked at my son aghast. 'Well, yes, that was going to be my question but how could you have possibly known I was going to ask you that?' Before we split up, his father and I used to joke that our boy may not be 100% human and could actually be from the planet Ork.

'I just knew,' Ryder said, grinning and returning to his native New England voice. During this time, his teeth were growing at alternate speeds and the top row looked like one side of a jigsaw puzzle piece. 'It's because the bottom of the leaves are silver.'

'That's right.' Anderson said, straightening himself up one silken vertebra at a time and gazing at me with his lichen-green eyes. 'You've got yourself a bright boy here, Ma'am.'

'Yes,' I said as Mr. Henry began to mock-fight with my son like they were twin brothers inside a play pen. 'It can be quite hard keeping up with him sometimes, that's for sure.'

I was looking forward to that batch of syrup but it never came. Poor Anderson just couldn't seem to get it right. His entropy kept going up and up and up and then his wife walked out and left him for a man twelve years her junior and there was nothing but more disorder all around him. Or so I thought until I caught him by the lake last Halloween tapping one of the silver maples with his small wooden peg. I bet Anderson had been talking it over with Copenhagen and saying things like: 'I'll make the syrup and then she'll come back to us. Don't you worry your pretty little fins anymore, Copen. Order will soon be restored.' But this didn't seem at all possible. Anderson was getting more and more frail; his syrup making was chaotic and utterly disorganized. Never mind Entropy Central, Mr.

Henry was living in Entropy City.

'Maybe it could be decreased,' my son interrupted. He was lying face down on our carpet, enveloped in the fabric of his new red sweatshirt.

'What was that, Ryder? Been reading my mind again, have you? Just how many of my thoughts did you actually intercept this time?'

'I got the one where you called Mrs. Henry a pass-hag.'

'I did not say, or think, pass-*hag*. I said pass-agg. It means passive aggress-'

'– Mom, I know what it means. I'm no *dummy*.'

'Please don't use that term, Ryder,' I said. 'And what exactly do you mean, "Maybe it could be decreased"?'

'Old dude Anderson's entropy,' Ryder said as he rolled over onto his back, and peered up at me with his fizzy cola-brown eyes. 'You don't know about the demon?'

'What demon?' I asked. 'You're the only gremlin around here, young man.'

Ryder let out a heavy sigh and stole a sideways look at his phone. 'Maxwell's Demon. He was able to sort shit –'

'– Language, please.'

'Sorry, I mean, sort *stuff* out.'

'Not sure I understand what you mean, honey,' I said. 'What sort of stuff, exactly?'

Checking his phone once again for messages, Ryder said, 'I gotta meet Jaden at the pool in like a half hour.'

'Have you finished your homework?' He shrugged as if to say, 'Duh. I did it in, like, 30 seconds.' 'Well, just give your old mum five minutes, at least. I'd like to hear about this demon.'

Another long exhalation of breath before my son at last began to enlighten me. 'So there was this Scottish dude in the 1860's or whatever and he came up with this theory. He wanted to know if it was possible to violate the second law of thermodynamics.'

My mind shut down completely for a moment. 'Yes,' I said, finally. 'I know that law, of course I do. Heat can only flow from a hot to a cold body and never the other way around.' He was staring at me, waiting for more. 'And entropy, or disorder, always increases in the end.'

While Ryder considered my reply, he wiped his nose with the sleeve of his new sweater. 'You sure about that, Mom?' He asked, inspecting the impressive assortment of snot.

'Use a handkerchief next time, Ryder, please. And yes, I am sure.' I was not at all sure. Ryder knew that I wasn't, either.

'But if you've got a demon that can, you know, get in there with the molecules and, like, infiltrate the system, then order might be restored?'

I still did not understand. 'But how? How would a demon, a mythical non-human being, be able to do such a thing?'

Ryder swept his thumb left to right across the screen of his phone. 'If I just tell you this one more thing, can I go?' I nodded yes. 'Say you have a container of gas with two parts. Let's call one part Ged and the other Ken –'

'– Wait. Don't you mean Jed and Ken?'

'No, Mom, just go with it. Ged an' Ken. Or you can say it all together: Gedanken. It's a German word and it means thinking or thought or something.'

He's teaching himself German now? Or did Ry just wake up this morning fluent in another language? 'I understand,' I said, lying. Ged an' Ken. Ged-an-ken. Gedanken. I repeated this word over and over. 'So, in this container are –'

'– Yeah, in there are, like, gazillions and gazillions of gas molecules, some are fast and some are slow and what this demon wants to do is sort out the molecules so that all the slow ones are in Ged and all the fast are in Ken and to do this he uses a –'

'– Whoa, just a minute, Ry.' He was talking at such a rapid pace, I could not keep up. 'Hold fast on those…story molecules of yours for a second, please?'

'O. K. Mom. I. Will. Try. To –'

'– Ryder,' I warned him. 'Any more of that sarcastic tone and you will not be going swimming at all.' He may be smart but I was still the boss around here.

My son sighed, brought the hood of his Hades-red sweater up over his head and finally began again. 'The demon has this trapdoor that only he can open. He knows when a fast molecule is getting close to the door so when it does, he opens it and it shoots through to Ged. And vice versa for the slow ones getting through to Ken. A gas' temperature is only an average speed of its molecules. Right, Mom?

'Yes...you are absolutely right, Ryder.'

'Even a hot gas has some lazy asses letting the side down and even a cold gas has a few speedballs. The demon would let these off-the-wall kinds of molecules through to the other side, which would mean the temperature difference between them gets even greater. Kaboom! That is it. The demon has just busted the second law of thermodynamics and caused heat to flow from a cold body to a hot one.' Ryder's soda popping eyes bore right through to mine. 'Hasn't he?'

Hasn't he? Why did I feel as if I was under surveillance right now from my own son? Was this a trick question? 'Yes,' I said. 'Or at least, I think he has.'

'Yeah, I guess. Maybe he has.'

'Maybe? But I thought you just said –'

'– Maybe he has and maybe he hasn't. Just think about it, Mom. Look, I gotta go,' Ryder said, sliding up onto his feet. 'But, you know, if he *has* then maybe Mr. Henry's ever increasing entropy can be reduced. Do. You. Under. Stand. What. I Am –'

'– OK, that's enough,' I said. 'You're free to go. And be back by ten-thirty,' I called out to him but he had already slithered away.

Old man Henry was always a cool buddy of mine. Mom had it right though; his entropy just kept going up and up. I didn't really get how such a super-ace dude like him could have been so sad about a pass-hag like Mrs. Henry leaving him but he was. Real sad. Mom said it was as if his heart was on the outside of his body and she could actually see it breaking more and more each day. My mom does have kind of an intense, over-active imagination but I got what she meant. She also said that the sight of him out there in his garden trying to boil up the syrup and failing every single time made her want to cry. And she did cry. A lot. Jaden's mom was always crying too, even when she just said hi to me. I don't get why moms do that. Anyway, I guess you might be surprised to find that I've taken over this story but I really had no choice. I couldn't keep on pretending to be my mom for too much longer so it's going to be me from here on in.

I really wanted to help out fine man Anderson. I think he was the oldest person that I ever met. I know he wasn't because that's always a chick in Japan. Sorry, *woman*. My mom doesn't like me saying 'chick.' 'Don't say "chick", Ryder,' she says in her weird accent, 'the word is "woman".' Before Mrs. Henry left Anderson, he didn't seem that ancient, but all of a sudden it was like he became 108 or something overnight. It was sad. I didn't cry about it but it did get to me in a really small, infinitesimal way. He was going to die soon. I knew it and my mom did too. That's why we wanted him to make the syrup so that he could at least go kind of peacefully and let Mrs. Mer Henry know what a good egg he was. So we came up with a plan. I was due to go stay with my dad for part of the Easter holidays and do you know what my dad's neighbour has in his garage? A reverse osmosis machine. This baby speeds up the whole sugaring thing. After sneaking into Anderson's place and lifting out his sap, it would mean that within a few days he'd have some of his own sweet-tasting Henry syrup. Yes, sir. Meanwhile, mom agreed to distract old Henry by inviting him and Jaden's mom, Clara, over here so they could all talk about the First World War or something. Just kidding, my mom was born in the eighties so she's not that past it; not as much as Jaden's who's from, like, the sixties. Anyway, that's what we decided to do to help out old man Henry.

We were all watching this documentary about salamanders at Entropy Central, I mean, at Mr. Henry's house when it happened. It wasn't really a big deal or a miracle or anything but I should probably tell you about it seeing as I'm the one who's in charge of the story now. My mom and I decided that this would be the night when we would steal Anderson's sap from his refrigerator. I figured that if we were all around at his place and it got kinda late then he would be asleep in no time, especially if we were going to watch a TV show about aquatic and semi-aquatic amphibians. It was actually pretty interesting but there was nothing in there that I didn't already know. Yet my mom, she just loves those kinds of things. She goes nuts whenever she sees one of the teensy orange ones (red-spotted but I already knew that) when she's out for a walk. She's got, like, hundreds of pictures of them. 'Oh gosh, Ryder,' she says. 'They're so gorgeous, aren't they? I can almost hear them chin-wagging with each other.' And then that crazy, bright, British brain of hers would be off again. Old man Henry was already starting to drop off before we'd even got to the part about some salamanders being able to regenerate their own limbs (totally the best part).

'Why don't you take Anderson upstairs to his room now, Ry?' My mom whispered to me from the sofa. I was lying on the floor, like I always do. Sofas are where all the old people like to sit.

'Yeah, if you want,' I said. I am a lot bigger than her so it made sense that I should be the transporter. 'Then you can go get his sap from the kitchen.'

'You got it,' said my mom. She almost made it sound like she was American but didn't really nail it.

As mom patted Anderson on his arm a little to let him know we would put him to bed, I went over to the TV to switch it off. Then I heard him mumble something like, 'Where are you, Meredith?' That was kinda sad.

'Are you OK?' My mom said to him. 'You feel a little hot. Would you like me to get you a cold flannel for your forehead?'

Old dude Anderson was burning up. Mom went to get a thermometer from the upstairs bathroom and after she returned and placed it in his dribbling mouth, the temperature read 101 degrees Fahrenheit.

'OK, Mom, you gotta go and open all the windows so we can get the air circulating,' I said.

'I know what to do, honey. I am a mum, remember,' she said. 'Let me get a flannel or a cold sponge.'

'Nah, that's not always so good –'

'– Well, it worked for you whenever you had a fever.'

'I never had any fevers. I'm a newt, remember? You must be thinking about one of your other kids.' She gave me one of her famous Medusa stares, then. She likes to do these every once in a while (my gran has an even better one, it's a British thing) and sometimes they can last about ten whole minutes. 'I'm kidding, I'm kidding,' I said. 'Maybe go get him a Popsicle or something like that?'

'I don't think Mr. Henry has any ice lollies in his freezer but let me just run home and pick one up from our house,' she said. 'I'll be back in two minutes.'

While my mom was gone, I broke my own no-sofa law and sat down next to our good neighbor. Mr. Henry took hold of my hand and held it. That was weird. I think he must have thought I was his wife because he kept murmuring, 'Meredith, Meredith' again and again. He felt real hot and his skin was so thin but also sort of rough that it kinda reminded me of one of those Pink Wafer cookies that I have whenever mom and I go visit her home town back in the UK.

'Hey, Mr. Henry,' I said. 'It's Little Newt. Remember me?' He didn't say anything for like an entire minute but just squeezed my hand harder.

'Ah, Newt,' he said eventually. 'There's my boy, my bright, special boy.'

'Yeah, that's right. You feeling any better?' No response. 'You want a glass of cold water?'

Anderson shook his head. 'Just sit with me a while, Little

Newt,' he said. 'Just sit with me.'

So I did. We sat together, holding each other's hands until my mom returned. She brought Clara back with her because we didn't have any Popsicles back at our place. I guess Jaden must have been at the pool or something which I was real glad about because if he'd have seen me holding old dude Henry's hand, I never would have heard the end of it on Twitter. They were both being all mom-like and fussing over Anderson, asking him how he felt and everything. Then my mom took his temperature again. It had risen to a hundred and three. She turned directly to me and said, 'Did you give him a hot water bottle or something?'

'No,' I said. 'We just sat here.'

'But he's even hotter,' she said before switching her touch from old man Anderson's hand to mine. 'And you're even colder.' My mom ran her fingers through her croppy, spiky hair. 'What have you done, Ry? You're making everything worse. Have you got tiny devils inside your pores?' No, of course I hadn't. But that's my mom for you; the simplest explanation is never true. I had done nothing, I swear, yet she and Clara practically yanked me up from the sofa and threw me out the door. I infiltrated my way back in of course because I had to get the sap; I had to make this goddamn Henry syrup if it was the last thing I ever did for old man Anderson.

The answer is no. For the demon to utterly defy the law, he would have had to sort out all of those molecules using, like, zero energy and zero work. But he couldn't have done. He had to have some super-knowledge about their movements and their positions to help get them into either Ged or into Ken. How would he have been able to open the door? Telekinesis? Don't crack me up. And you think I could have just made Mr. Henry hotter purely through the touch of my skin? A coincidence; pure and simple. You can't cheat the principles of nature but you sure can have fun trying.

For a while, it seemed as if my mom and me had cheated the death out of old Mr. Henry. We made jars and jars; no, I am talking casks, barrels and vats of his maple syrup. We literally lined his hospital room full of the stuff. It did get kind of dangerous because a few of the nurses even tripped over them but they weren't seriously hurt. Anderson's last few weeks were pretty good ones and he talked a lot about he and Mrs. Henry getting back together again. Mom even helped him write her an email but she only wrote back saying: *I loved you, Anderson. I really did. It just became wearisome being thought of as the Wicked Old Witch married to the mature Prince Charming. Did you ever really love me as much as you loved our neighborhood?* At least she included a question mark this time, yet that mark pretty much sent my mom over the edge. She kept saying how she must try to curb her fiendish imagination, that she had possibly misjudged Anderson's wife all this time and Meredith really wasn't so bad. I don't know. Mer Henry had almost smiled at me once, though she might have been trying to get something out of her teeth.

From there, everything else started to fall apart for our good neighbor again. Henry's goldfish, Copenhagen, was found dead in his tank, but goldfishes never live that long. I don't know why anyone bothers with them. They're kind of pointless so he really shouldn't have taken that death so personally. Yet when I said this to my mom, she just bawled and bawled. And then when poor old Anderson Henry actually died, she cried for, like, three whole days. Jaden and me had to take our sleeping rolls and go camp out in the woods just to get away from our crying moms.

In the end, the arrow of time only points one way for all of us. People get old. People get sick. People die. But the Henry syrup? That's still going strong. My mom and I were drizzling some of it over our breakfast waffles just the other day. I think it is so awesome how it sinks into the square voids of the waffle so you can't see the gaps anymore. It looks real ordered and uniform. Although do you know what my mom did then? She took hold of my fork and smashed into it so that the syrup spilled out all over the plate and the waffle just got utterly demolished. Then she started mimicking my voice and saying things like, 'Before I destroyed it with a fork, this waffle was low entropy and now it is high entropy – kaboom!' She did manage to pull off how I sound real well, actually.

But then she has been getting better at pretending to be me.

Afterword:

Maxwell's Demon

Dr. Rob Appleby
University of Manchester

PROBABLY ONE OF THE more famous of the thought experiments, James Clerk Maxwell's imagined demon is an attempt to directly violate the laws of thermodynamics though a mischievous demon between two boxes of gas.

The laws of thermodynamics were formulated in the 1800s and are, even now, important pillars of the modern scientist's view of the world. The laws tell us about how heat works and provides an arrow to the flow of time. The zeroth law states that if two systems are separately in thermal contact with a third system and no heat flows, then no heat will flow if the two systems are in thermal contact with each other. This sounds trivial, and it is, but helps us make sense of ideas like temperature and temperature scales like 'degrees Celsius'. So we find it useful to write down a seemingly simple statement. This zeroth law helps define something important called 'thermal equilibrium', which is what two systems are in when they are connected but no heat flows between them. Happily, the first law is more interesting and essentially says that energy is conserved. More precisely, it says that if I heat a system or push on it (do work to it) then that system's internal bookkeeping measure (recording how much energy is contained) must show an increase by exactly the right amount. It also says that how the heat is applied to the system, or how the work is done to it, does not matter. Only the total amount matters in the end. There's a nice equation that comes with the law that lets us understand phenomena like compression of a gas, heat capacities of substances or fluid flow. Next we come to the second law, which has by far and away the most number of forms and is, by far and away, the most interesting. Imagine we only had the zeroth and first laws. Then heat could flow from one body to the next but there is nothing to say which way the heat flows. This means heat could flow from a cold body like an ice-cube to a hot body like a cup of tea and not violate the laws of thermodynamics. However, if we look at nature we see a natural direction for change – heat flows from a hot thing to a cold thing and gives a sense of direction. The second law fixes this by making a statement about the irreversibility of nature. In the (paraphrased) words of German physicist Rudolf Clausius:

> 'No process is possible whose sole result is the transfer of heat from a colder to a hotter body'[20]

So heat flows from a cup of tea to an ice-cube. There is also a more intriguing formulation of the law, involving entropy. 'Entropy' is the scientific term used to describe disorder and the fact that we know the universe likes to move to a more disordered state. A trivial example is the aforementioned tea cup which can fall and break into many pieces but tends not to spontaneously unbreak and form a perfect tea cup from the many little pieces. More precisely, the second law of thermodynamics states that the entropy (disorder) of an isolated system cannot decrease. This immediately provides an arrow of time and means systems tend to get more disordered with time.

What has this got to do with Maxwell? He was working, shortly after the laws were formulated, on a pioneering microscopic theory of gases and wanted to explicitly pick a hole in the second law and show it only had a kind of statistical certainty. To do this he imagined a demon that was able, by controlling a little window between a box of hot gas and a box of cold gas, to effectively make heat flow from the cold gas to the hot gas. This was done by preferentially opening the window when 'cold' gas molecules (on the hot side) headed towards the window, letting them through to the cold side, and vice versa with 'hot' molecules on the cold side. This would make the hot side hotter, and the cold side colder, directly contravening the second law and preventing the inevitable rise of entropy.[21]

It took almost 100 years for the resolution to this thought experiment to be discovered. The first counter-argument came from the act of measurement itself, and the fact that the demon would have to expend energy to measure the gas molecules as they came near – thus increasing entropy. This argument stood until it was pointed out that, in principle, the demon could make a measurement without expending energy but at the cost of storing all the information gathered during the measurement process. The demon, presumably being a finite being, would not have infinite storage and eventually would have to erase old data to make way for new. These are the beginnings of arguments founded on information theory, which are of central importance today in fields such as computation and quantum mechanics, and with them Maxwell's demon was finally dispelled.

Notes

20. First expressed in his paper, *On the Moving Force of Heat*. See Clausius, Rudolph (1867). *The Mechanical Theory of Heat – with its Applications to the Steam Engine and to Physical Properties of Bodies*. London: John van Voorst.

21. The thought experiment first appeared in a letter Maxwell wrote to Peter Guthrie Tait, 11 Dec 1867, then again in a letter to John William Strutt in 1871, before appearing in full in his book *Theory of Heat*. Maxwell called it a 'finite being', but it was William Thomson (Lord Kelvin) who first christened it a 'demon' (*Nature*, 1874).

The Rooms

Annie Clarkson

Hannah felt a burn in her thighs from pedalling too hard. She was late. She had mismatched socks, and her hair was held back with a roughly tied headband. She was starting another work trial today. This was the nearest she had been to full-time employment since being made redundant a year ago. The pay was way below what she was worth, but she needed the job, even if she didn't fully agree with what it would involve.

Her tyres spray-patterned the backs of her legs with mud. Lamps glared and faces stared onto the canal from the windows of converted mills. Hannah skirted the railings onto Ancoats Street, turned right, left, and round the back of some warehouses there was a new building with a red door, just as Alex had described.

She pressed the buzzer, trying to smooth her unruly hair. The door beeped and the voice on the intercom said, 'Fourth floor'. There was no lift, nor any doors on any of the stair landings, just fake plants in huge pots and arrows next to tiny signs that said 'The Rooms'. Her stomach tightened a little, as she realised Alex hadn't told her who to ask for when she arrived.

In her mind, she imagined a receptionist looking at her blankly when she said her name. But, there was only an unmanned desk and a screen where it was apparent she needed to sign in. An ID badge was automatically printed and delivered to her from a slot, along with a confidentiality contract that she had to read. Hannah pressed her thumb against the screen, as confirmation that she agreed not to discuss her work with any non-employee and a door slid open.

In the adjoining room, there were framed prints of lotus flowers on the wall and a gold Buddha next to a water feature that hummed and trickled in the corner. There were five low chairs and a man wearing an expensive tailored suit. He was drinking water, not sips, but huge thirsty gulps until the glass was empty. He nodded at Hannah and smiled at a petite woman who came into the room, bringing his coat. 'Thank you,' he said. 'See you same time next week.'

The woman reached out a hand for Hannah's coat. 'You will have nice time,' the woman said, which struck Hannah as odd, but she didn't have the chance to respond before Alex came into the room.

He grinned. 'Hello Sis.'

'Sorry I'm late.'

He shrugged. 'It's fine, come on, let me show you around.'

She followed him down corridors lined with closed doors, each with a number on. It seemed like a maze. At each turn of a corridor and each junction there were signs, as in a hotel, showing which direction to take. She tried to keep up with his instructions, 'Each of the girls is assigned a room. They're always in the same room, always the same work. The clients are assigned to a girl that matches their specifications. Your job is to support them, to improve their communication. I've matched you with Myla. She's one of our higher spec girls: highly natural, self-learning, empathetic. She brings in our best clients. She'll train you, and once you're ready, if you *are* ready, we'll agree a contract and you can start with the other girls.'

'Are none of them male? Like, is the whole place just full of females?'

Alex laughed. 'Sis, there isn't much call for male bots in this field. Think about it.'

He showed her how to swipe her ID card against a panel outside the doors, and almost pushed her into Room 43, a hand on the small of her back, as he shut the door and left.

It was small, no windows, only a dim light from a lamp on

a bedside table. There was a neatly made bed, and a woman sitting at a desk facing away from her. The woman was dressed in a business suit. Hannah wished she had made more effort with her appearance. She smoothed down a lick of hair that kept sticking out at the back, and cleared her throat. 'Good morning.' The woman swivelled round in the chair, and Hannah's mouth dropped open, before she realised she was gawping and quickly tried to hide it with a smile.

Myla was beautiful and unlike any person Hannah had ever seen: perfect straight white teeth, perfect almond eyes, full-pouting lips. Her skin was olive and smooth, without blemish. Her hair was sleek, trimmed to shoulder length, not a hair out of place. Hannah could see no fault in her whatsoever, and realising this made her feel conscious of her own freckled skin and ever so slightly crooked teeth. She chided herself for the comparison.

Myla pushed a folder towards her and asked her to read through the day's tasks before they began. She started to read, her hands trembling slightly. Hannah's eyes scanned the list of instructions and questions, but she struggled to focus.

Myla smiled at her kindly. 'You expected me to be less realistic?'

Hannah nodded, although she wasn't sure whether Myla was realistic. She was too perfect. And her voice had a strange inflection to it, almost an underlying purr, not mechanical as such, but like the humming of a radio on standby. She spoke in pure Queen's English, no sign of a Manchester accent. Hannah could see how she brought in the best clients.

Hannah's job was to talk with Myla using the subject areas indicated in her daily task sheets. There was a list of questions she could ask, with variants, and sub-topics. There was also a list of desirable answers, which Hannah was to guide Myla to respond with, and answers that she had to indicate were not appropriate.

Hannah rushed ahead, thinking that she would soon get through the topics, but Myla kept stalling. It was obvious Myla was deliberately using inappropriate responses to see if Hannah would follow the instructions exactly, and when Hannah's patience almost slipped a couple of times, Myla said they should take a break.

'Is there somewhere to get coffee?' Hannah asked, 'and a bathroom?'

Myla highlighted a screen next to the door where Hannah could choose from a list of foods and drinks, to be delivered to the room. Myla pressed a button and a door slid open to an ensuite toilet and wash area.

'So, where's the staff room?' she said. The size of this building, she figured there had to be a cafe or staff room somewhere, to sit and chat to colleagues in, read a magazine, look out of a window.

Myla smiled, 'Food and drink is brought to the room, and the bathrooms are private. It's policy to remain within your allocated room.'

Hannah frowned. It wasn't what she expected. It was only 11:08 and already the confinement of Room 43 was making her feel uncomfortable.

She was relieved at the end of the day when the door released and Alex stood waiting for her. She followed him out to the waiting room where the petite woman with immaculate make-up held out her coat and said, 'See you next time.' Hannah fastened her coat and glanced at the two men in the waiting room, one reading a newspaper, the other sitting with his eyes closed and rubbing his fingers against his temple as though he had a headache. There was low music and a dim blue light from the water feature. Neither of the men looked in her direction or acknowledged her. She wondered who they were, what kind of men they were to be here. But, Alex hurried her out, into the unmanned reception, where she swiped her card and watched the letters of her name dance across the screen with the words, 'Goodbye for now'.

Hannah felt deflated. As she cycled back along the towpath, with the setting sun pinking the mill windows, she wondered whether she was going to last even the trial period. The day had seemed so long. There was nothing to look at other than the four walls and perfect Myla and that neatly made bed.

In the whole seven hours they'd spent with that bed taking up half the space in the room, they hadn't once mentioned it: the reality of her job right there in the room.

'A job is a job', Alex had said at the time, and she had struggled to argue against him, as he was letting her stay in his house and had been providing for her since her savings had run out.

Back at the house, Hannah threw her bag at the bottom of the stairs, kicked her shoes off and went to make herself some coffee. She decided she wouldn't look at the men in the waiting room any more. She would just follow her tasks. Alex was right, it was a job. She was having conversations and those conversations would mean she would get paid. That was all.

Each morning, Hannah found her way to Room 43 and swiped herself in. Sometimes when she arrived Myla was lying on the bed with a wire connecting her arm to a socket in the wall. The first time, Hannah felt as though she was intruding. She fumbled with her ID card and tried to swipe herself back out, but the door was locked.

'Don't worry, I am almost charged.' Myla's eyes flickered open and she smiled at Hannah. 'It's my equivalent of breakfast,' she said, 'Have you eaten?'

Hannah had skipped breakfast, and actually, felt quite hungry. 'Order something,' Myla said. 'It has been a difficult few weeks. I pushed you to work harder than I should have. Maybe we should take it slower today and have more breaks. I forget that you get tired.'

'Do you not get tired?' Hannah asked, and then felt stupid for asking. Of course, she didn't get tired. But Myla smiled and said, 'We have our own kind of tired, but it comes on suddenly, if we don't have enough energy', and Hannah laughed and said, 'I'm like that sometimes, as if I run out of batteries.' They both laughed.

Hannah ordered toast and fruit from the screen beside the door and it was delivered about five minutes later by a tray on wheels that brought itself to the door. She read her tasks and started on the list of questions about likes and dislikes which was today's topic:

'What do you like about your job?' she asked, 'Is there anything you dislike?', and then, 'What do you like about me?'

Myla listened and responded, and sometimes she would repeat what Hannah said. It had taken some time for Hannah to realise that Myla often mirrored her: the way Hannah moved her hair to one side and tucked it behind her ear; the way she bit her lip as if she were unsure; the way she paused and blinked and cleared her throat, and sometimes just paused to think.

When Hannah realised Myla was copying her, she crossed her arms and Myla crossed her arms and they stared at each other until Hannah couldn't stand the silence any more. 'Why are you copying me?'

'It's the task today.'

'That's not what it said on the instructions,' Hannah said.

'It was in my instructions,' Myla said.

'You have instructions?'

'Of course.'

Hannah was silent again. She had thought she was in control, that she was asking the questions, that she was the teacher, but suddenly it felt as though Myla was the teacher and she was a student and it made her feel uncomfortable.

'I need a break,' she said. Hannah tried to swipe her ID card against the door lock screen, but it didn't work. She went into the bathroom instead and splashed water onto her face. She counted the length of her in-breath and then her out-breath. She waited until she had completed ten full cycles of breath and then opened the door.

Myla was sitting at the desk. Hannah sat down on the opposite chair ready to start her job again, but the topics on the cards were not the ones she wanted to talk about. She stared at the list of topics, and put the card down.

'Why were you copying me?'

'Why am I copying you? Part of the training is to copy.'

'Why?'

'Why? Alex says that he wants us to develop character.'

Hannah wondered if Myla even knew what character was. Myla began every answer by repeating what Hannah had asked, using the same inflection, tone and rhythm. She paused when Hannah paused. Without any effort, she seemed to adopt every gesture of Hannah's, even the slightest flicker of an eye or clenching of the jaw. It made Hannah self-conscious and by the end of the day she wasn't sure which of Myla's responses were genuine and which were copied.

Hannah left the building exhausted. She decided to leave her bike and get the bus home. She found the nearest seat to the front and sank into it, so that her eyes were level with the bottom of the window. She couldn't focus on anything, and the colours of passing cars were just shade and light, colour bleeding into colour. Her thoughts were still in Room 43 with its plain pale walls, the low-lighting that made it feel as though the ceiling was pressing down on them, the screen by the door which faded to blue with a clock flicking from minute to minute through the day as she asked endless, and it seemed, pointless questions.

She walked in streetlight from the bus-stop to their estate, as her brother passed in his car, not stopping to pick her up. He had already parked in the drive by the time she reached the house. She found him in the kitchen making a smoothie with beetroot and spinach. The whirr of the blender made it difficult to talk.

'What is my job exactly?' Hannah asked. She had to repeat it louder and then waited until the blender had finished and asked again.

'I heard you,' he said, drinking his smoothie. 'It's a job Hannah. It's doing whatever you are told to do and getting paid for it.'

'But what's the purpose? I go to that room every day and repeat myself and talk round and round in circles and she copies me and responds to me, and I tell her if I think she's right and wrong, but she can do everything anyway.'

'It's a job.'

'You already said.' Hannah picked at the paint on the edge of the kitchen door where it had dried unevenly. 'It's frustrating.'

'But you're doing great,' he said. He grabbed his running shoes from behind the door, and said, 'I have to go kid, I'm back in for a late shift tonight. Listen, let's talk about it at the weekend. Come out for a drink with some of the others.'

Hannah shrugged, indicating maybe. She hadn't met anyone else who worked there, only the woman in reception who took coats and brought glasses of water to the men who visited The Rooms. She didn't pass anyone in the corridors. There was never any contact with anyone, other than Myla. There were all those closed doors and she assumed behind those doors were other women and other employees like her.

At the weekend, Hannah met Alex in a pub near their offices. It was a retro place, a bar with real fires, and various rooms to sit in depending on whether you wanted a quiet room or a room with music playing. There were tiles on the walls, stained wood on the floor and cracks in the ceiling, but it was warm and Hannah liked the cosy feel. It was unlike the stark stripped-bare places she imagined Alex would like. There were trinkets everywhere, pictures hanging from strange hooks and glasses shaped like tankards.

She found Alex in a back room with three other men, drinking beer, and playing cards. She squeezed past people, apologising as there wasn't much room for people with all the furniture.

'My sister Hannah,' Alex said to the men, 'our newest recruit.' The guys all nodded and one by one shook her hand. Michael with the long fringe, Jav with the curls, and Samuel, who didn't look Hannah in the eye, but smiled in a shy way, and carried on laying his next card down. 'What you playing?' she asked. 'It's not a game,' Jav said. 'It's work, I'm afraid.' He shifted along a wooden bench to make room, and Michael asked whether she wanted a drink. Hannah asked for a beer. Beer was not really her drink, but she felt it made her one of the group if she drank the same.

'Is it largely men that work at The Rooms then?' she asked, hoping some women would be along soon.

'Yep, all men,' said Jav. 'You're the first woman to come and work with us.'

'What about the woman from reception?'

They laughed, 'Nope.'

'Oh,' Hannah realised what they were saying. 'So, I'm the only woman?'

'Yep, it's an experiment, of sorts,' said Jav.

'Not an experiment,' said Alex, and they had a small discussion about whether it was or wasn't an experiment, and really how it was a trial or a test, but it was going well, they all agreed, it was going far better than expected. Even Alex had to agree that it was a surprise how positive an impact Hannah was having on the business. They nodded and smiled, and even the shy one looked at her then. They all said, 'Here's to Hannah,' and raised their glasses, although Hannah didn't have a glass to raise yet, and Michael wasn't there to raise his glass, but they raised their glasses anyway and Hannah felt they were almost congratulating themselves.

Hannah was cramped up to Jav on the bench and had to try to ease her jacket off without elbowing him or bumping shoulders. It was hot in the small room with a fire burning in the grate, and the close proximity.

'So, what are the cards?' She asked.

Samuel pushed one across the table towards her, and she saw that, in tiny red writing across the back of the card, was written over and over 'The Rooms.' She turned it round and there was a picture of a woman on the other side. Michael squeezed his way back to the table and sat opposite her. 'You picked Nina,' he said. 'Good choice.' Hannah looked again at the card, at the photograph of the woman who was very young, long haired, dark skinned, beautiful, and flawless like Myla. Hannah reached for the pack and took some more cards, looking at the different women on the cards, flicking through the pack.

'How many are there? She asked.

Jav laughed in an excited way, 'Eighteen!' he said. 'Can you believe it? We're developing five more at the moment. With your help they will prove far better than any of the others. Tell her, Alex.'

Alex laughed in more of a nervous way, 'One step at a time, guys,' he said. 'I want Han to settle in first. She's still getting used to things.'

Hannah felt cramped again and the heat. There seemed to be so much they weren't saying, so much they could have said, but it was all left unspoken on the table with the photographs of women in front of them.

Hannah decided to finish her drink and tell them she had to leave to meet friends. She was lying of course, and it was difficult to make an excuse when they were so friendly to her. Michael asked questions about her life and seemed to want to make her feel part of the group. Samuel was quieter, and with a sheepish grin he kept apologising every time he checked his phone and sent messages. Jav wanted to tell her about the second generation design and the software they were using, how as well as being self-learning, this version would be able to replicate emotions in ways that the first generation girls couldn't.

It was hard not to share their enthusiasm, but Hannah felt as though she was a guest at an event she hadn't quite understood.

It wasn't until the following week, after Hannah had finished her probation period, that she started to visit other rooms. Identical rooms, with different women, who were all different versions of perfect. Except Alina.

Alina was willowy with pale creamy skin and wrists so slight they made Hannah think of twigs. Her hair was fine blond and fell in baby curls against her neck, stopping short of her shoulders. Alina smiled. She moved a strand of hair to one side and tucked it behind her ear. She bit her lip as if she was unsure of Hannah.

Hannah found her instructions on the desk and started the day's tasks. The first topic was related to memories and how to answer questions about them.

'What did we talk about last week?' Hannah asked.

Alina didn't respond.

Hannah tried again. 'Can you remember what we talked about?'

Hannah pulled a chair over to where Alina was sitting on the edge of the bed. She asked Alina questions and variants on those questions, but although Alina's grey eyes watched Hannah and followed her movements, she didn't speak.

Hannah didn't want to get into trouble for not doing her job. She asked more questions and waited for Alina to perhaps shake or nod her head, give a gesture or a sign of some kind. But Alina wasn't communicating and that confused Hannah. She'd seen no sign of this from Myla, so perhaps Alina wasn't functioning properly.

Hannah kicked off her trainers and sat at the desk deciding she would read a magazine for a while. It was distracting with Alina in sight, perched on the bed, eyes blinking wide and staring. She tried to focus on the page, but her concentration wandered, and even though she knew that Alina was a machine, in a way, she couldn't help but feel concerned for her.

Hannah asked if there was anything Alina needed. She went to sit next to Alina on the bed, and asked her what the matter was, why was she so quiet, was there anything Hannah could do? Alina leaned against Hannah's side and slowly closed the space between them until Hannah found herself with her arm around Alina. Hannah was quiet when she spoke. She suggested that Alina lie down for a while, and helped her to pull back the covers as Alina curled up on her side, and closed her eyes.

The day passed slowly, with Hannah ordering snacks and drinks, and idling around the room, searching for drawers or any hidden spaces in the room. She checked on Alina regularly, who responded when she spoke, but didn't seem to want to engage.

Near the end of her shift, the door slid open, and Hannah found Alex and Jav waiting in the corridor outside. 'How was she?' Alex asked.

Hannah said, 'She was tired.'

Jav laughed, 'That's impossible.'

'I didn't have my instructions today,' she said, frowning and fastening her coat ready to leave.

'Don't be annoyed Han,' Alex said, and put his hand on her arm. 'Come on, meet us for a drink, Samuel wants to ask you about Alina. There's something interesting happening we don't understand.'

Hannah agreed, although she was tired from a day sitting around doing very little. She made her way to the waiting room. There were two men, one of them about to leave. The petite woman brought his coat. She smiled and said, 'See you next time.' The woman nodded to the other man. 'You are here for Alina.' Hannah stared at him. He was maybe fifty, with a receding hair line that he tried to hide by keeping his hair shaved. He was short and heavy. Hannah couldn't tell what kind of man he was. He was just ordinary; could have been anyone, working anywhere, someone's friend, husband, dad.

Hannah said to the woman, 'But Alina is not well.'

The woman just smiled and turned to the man, 'You will have a nice time,' and showed the man through into the corridor.

Hannah ran down the four flights of stairs and onto the street outside. It was cold, and she had forgotten to collect her coat. She pulled the sleeves of her jumper down over her hands and hugged her arms to her body. She wanted to go back upstairs to Room 35 and bring Alina outside with her, so they could go somewhere, anywhere, get a drink and talk, away from The Rooms. But, at the same time, she knew. None of the women left their rooms or the building. They didn't go out for drinks or talk about everyday things. They were designed to do a job and they didn't need rescuing.

Hannah walked to the pub, but she dawdled, and by the time she arrived, her hands and nose and chin were pinched with cold. She went into the smallest room in the pub, and sat on a table near the fire thinking about Alina. She was different from the other women. It was like she had made a decision not to work on the tasks that day.

She left the warmth of the fire to get a drink from the bar. Samuel was already there, shifting from one foot to another, clearly cold. He smiled in his usual shy way, making eye contact but looking away again, as though he found human interaction awkward.

'Alex will be along soon,' he said, and she smiled, feeling his awkwardness spread to herself, as his eyes met hers and looked at the floor, then the glasses along the bar top and then her eyes again.

They sat at the table near the fire and Hannah wanted to ask questions, but wasn't sure how to word them or in what order, or even where to start. Finally, after a long silence where they each stared at the fire, she asked him, 'What's wrong with Alina?'

He frowned and said, 'We don't know.'

'It's like she's depressed.'

He shook his head. 'No, we don't programme them to have feelings, only to replicate behaviour and Alina has never seen that kind of behaviour to replicate it.'

'But maybe she learnt to have feelings? Maybe she learnt to understand what feelings are.'

He shook his head. 'No, that's not what we programmed.' Samuel bit his lip. 'She functions really well most of the time,' he said. 'She carries out her job well. We don't understand why she's inconsistent.'

'It's like she was just so worn out today that she didn't want to work.'

'It's easy to project human feelings onto the girls,' he said. 'Alex believes this is why it's successful, people believe they are real, that they live and breathe and feel like us. But it's not the case.'

'How do you know?'

Samuel leant over his pint, and stared into it. He got some papers out of an inside pocket and laid them out, showing her data that she didn't understand. Lines and lines of data, which seemed to make him more animated, as he pointed out this and that. Hannah didn't follow what he was saying, but she followed the cadence of his voice, as it rose and fell in a way she hadn't imagined him capable of. He told her that Alina was a second generation girl, how they developed the main programming, but were struggling with some aspects of the design, and he hoped Hannah could help him understand, given that the second generation girls were modelled on her.

He was watching her carefully, his eyes still darting a little anxiously, but with all his focus on her. He looked from her eyes to her fingers tapping on the table to her bitten lip to the strand of hair falling in front of her eyes that she couldn't brush away because she was still trying to take in what he had just said.

She cleared her throat and said, 'In what way, modelled on me?'

'Oh mannerisms, body language, inflections in your speech, the way you interact and communicate. It's the idiosyncrasies of a person that were missing from the first generation. We made her too perfect basically, and that's not what some people want; it makes them feel inadequate. We learnt a lot from you when you were working with Myla.'

Hannah wasn't sure how to respond. It was clear Samuel assumed she knew this already. She frowned and asked, 'You were watching me with Myla?'

His eyes dipped to the floor and he mumbled a few words that she couldn't make out.

Hannah wanted to shout at him, but really, she knew it was Alex who needed to answer her questions. Alex, who had failed to tell her exactly what her job role was, evaded questions, and told her, 'A job is a job'. She thought back to when Alex had started the business, and how little he had shared about it. Even now she was working there, everything she knew about the business was guess work and assumptions based on what she saw and felt and heard.

She stood up suddenly, startling Samuel, who then began gathering the papers from the table, and stuffing them back untidily into his inner pocket. He stood up too, pushing the barstool back with a screech against the wooden floor and then apologised. She wasn't sure whether it was an apology for the noise, or for his realisation that she was not responding in the way he expected, and this meant he had told her some information she wasn't happy with.

'I need some air,' she said.

'Should I come?' he asked, and she shrugged and walked towards the door with him following. They stood in the street outside, with its mismatched paving and row of thin saplings bare of leaves.

'I'm sorry', he said, and she shrugged again. It wasn't him. It was Alex, typical Alex, always getting his own way, leading her into situations that she didn't understand.

Samuel walked her home and they talked about the moon cycle, currently a day or two short of a new moon with only a sliver visible. They talked about the Manchester skyline with its chimneys and cranes and criss-crossing electricity wires, and how it had changed with the new builds, and was always changing. They talked of trams and the history of the streets they walked and the rain and their favourite cafes around the city and anything except The Rooms and the girls there. They skirted around the subject and somehow talked more freely because of it, none of the restraints of previous conversations.

When they reached the house, Samuel said goodbye and sorry again, and left before she had put her key in the door.

She didn't sleep well, instead spent most of the night thinking about Myla and Alina, the men in the waiting room, her task lists, and her own options right now, which were not many if she were honest. Work was hard to get and the past year had been interview after interview, failed applications, and useless work trials. She needed this job, even if it was only until she found another.

The next morning she walked along the canal bank. Ice cracked in the puddles as she walked, and pigeons on the path waited until she was one step away before noisily flapping up and away. She didn't know which girl she would work with today – Myla or Alina or another girl –but she would have her task list, and she would follow it, and the girls would learn different ways to respond to a question, and what it meant to furrow their brow, or yawn, or turn their back on a person when they were speaking. She would teach them these things and others. She would teach them to be honest - even if that meant being difficult - and she would teach them to say no sometimes.

Afterword:

The Chinese Room

Prof. Seth Bullock

University of Bristol

> John: 'OK, Google.'
> Google >> [boop]
> John: 'Do. You. Understand. Me?'
> Google >> Does anyone really understand you, John?
> John: '…'
> Google >> …

In 1980, when philosopher John Searle originally published his Chinese Room thought experiment, he might not have imagined that within his lifetime his own phone would be able to correct his spelling, anticipate his next written word, translate his ideas into many languages, and, perhaps most impressively, answer his spoken questions, instantaneously and in some cases quite accurately. However, back then, at a time when people still thought digital watches were a pretty neat idea, Searle had already determined that no matter how clever a computer might box, its smarts would always be a sham. His thought experiment made clear that, regardless of how brilliant his phone was, the question of whether it could be said to *understand* his questions (or its own answers) was a different and far more vexed matter – because, according to Searle, the outward appearance of intelligence need have nothing to do with real understanding, experience, or meaning.

By 1980, artificial intelligence (AI) practitioners had got used to the idea that the gold standard for an intelligent algorithm was passing the *Turing Test*: behaving in a way that was indistinguishable from a real person. Thirty years previously, while Alan Turing wasn't winning World War II by cracking German cyphers, or laying the theoretical foundations for the whole of computer science, or explaining how the leopard gets its spots while the zebra gets its stripes, he had suggested that (in the absence of agreed criteria for what thinking actually amounted to) it made sense to adopt a kind of 'duck test': 'If it looks like a duck, swims like a duck, and quacks like a duck, then it probably is a duck.' Analogously, he suggested, while we might not be able to define what it is to think, like pornography, we would surely know it when we saw it. Searle set about showing that this could not be true: if a thinking thing had the wrong stuff inside it, then it could not really be thinking. And, heretically, for Searle, the 'wrong stuff' turned out to be: computation, logic, symbols, calculation. The very bedrock of AI and computer science. But surely, his readers balked, thinking *was* computation? That had already been agreed? The Chinese Room must be a trick, a riddle. Like a locked-room mystery, we would have to think our way out of it.

In his thought experiment, Searle imagines himself as a kind of isolated slave-clerk, working alone in a room with a desk, a chair, a pen, an in-tray (into which new work for him arrives), and an out-tray (from which the products of his labours are removed for use outside the room). He also has a book of rules that he must follow. Written in English, they instruct him to take sheets of paper from his in-tray (which are covered in unfamiliar 'squiggles and squoggles') and perform operations on them. Exactly which rules must be followed and in which order depends on which squiggles and squoggles are on the incoming sheets of paper. Sometimes Searle must cross-reference incoming symbols with those in the rule book, or copy incoming symbols into specific places in the rule book, or sometimes erase or alter parts of the rule book. But ultimately, the rules require that he add his own specific squiggles and squoggles to a new sheet of paper and place it in the out-tray. During this activity, Searle fully understands the rules that he is following – they are written clearly and explicitly in English which is his mother tongue – but he has no additional knowledge or understanding of the job at hand.

Now, unbeknownst to Searle, written on the incoming sheets of paper are questions or instructions in Chinese (a language that Searle has no knowledge of) – this is what the squiggles and squoggles are. And, also unbeknownst to Searle, the outgoing sheets of paper that he has worked to create are actually responses also written in Chinese. In fact, it appears to Chinese language users outside the room that they are having a normal conversation with a Chinese language user in the room.

Searle's claim is that regardless of how successful this Chinese conversation is, there is no understanding of Chinese inside the room. Searle-in-the-room does not understand Chinese. He has no experience of the meaning of the questions or answers, or his own activity. The Chinese Room is just meaningless symbol processing, and there are therefore no grounds for believing that a *computer* carrying out the symbol processing would understand what it was doing either. Since all that a computer *is*, is a machine that carries out symbol processing instructions, computers are evidently not the right type of things for doing thinking, understanding, meaning, comprehending, appreciating, fathoming or grasping, no matter how convincingly intelligent their behaviour might be.

The reaction of Searle's readers was stark: Searle is wrong. Wrong about thinking, wrong about computers, wrong about symbols, wrong about the world, wrong, wrong, wrong. But Searle had spent the months prior to writing his paper touring philosophy departments and AI research labs, presenting his thought experiment, and when he came to write it up, he helpfully included a list of the major objections to his idea. And quickly dispatched each of them. Maybe the whole room understood Chinese even if the man inside didn't? Maybe the man in the room did understand Chinese, but just didn't realise? What if the room were actually the head of a robot, and controlled its behaviour in the real world?

Different readers reacted differently to this parade of attacks and rebuttals, finding some of them compelling or unanswerable, some unconvincing or even risible – depending on their theoretical position, but also their assumptions, prejudices, and ideology. But for any reader, regardless of the conclusion that they reached, something was inescapable: it was no longer easy to equate the processing of information in a machine with 'thinking' in the sense that we normally use the word. Searle had at the very least convinced us that there was a problem that had to be tackled.

For Searle, it was perplexing that anyone had ever thought that computation could be thinking: 'The computer has a syntax [rules] but no semantics [meaning].' He pointed out that no-one would expect a beautiful waterfall screensaver to actually be wet, or that a simulation of a fire could 'burn the neighbourhood down'. No matter how realistically coded, the machine just doesn't have the right 'causal properties' for wetness or burning. Likewise, a computer mimicking the way that a brain works, will never actually have a mind (the very idea!) and consequently will never think.

As a somewhat nihilistic teenager encountering the Chinese Room a quarter of a century ago, I was taken with one possible reading of the Chinese Room: maybe true thinking, meaning, understanding just doesn't exist at all! Our wet squishy brains *are* just computers, and Searle is right that while computers can give the appearance of all of these things, they cannot truly possess them in the way that we had assumed. Like a pulp sci-fi movie, the hero discovers that all the people around her are really just soulless automata, and, in the final scene, she manages to loosen her own mark to discover that she herself is also inhuman! Thrilling stuff, but, I came to realise, ultimately unsatisfying.

Returning to the Chinese Room these days as a middle-aged academic drone, I (perhaps predictably) find myself more likely to dwell on the plight of Searle-in-the-room himself. Locked inside on his own, meaninglessly processing 'work' that requires no insight, skill, or humanity, disconnected from the real world, and at risk of being replaced by a machine. A rather modern-sounding predicament, but there is a strong resonance here with the original 17th century use of the word 'computer': a person employed to carry out simple calculations that contributed to some larger enterprise in a way that was unlikely to be meaningful to the computer themselves.

So where might meaning come from in life and in work, if not from symbols, rules and computation? There are many responses, but most share the same basic answer: from the world. Symbols need to be *grounded*. While the meaning of a symbol can be partially glimpsed from examining its relationships with other symbols (the symbol *cars* means 'more than one car', a 'car' is a kind of 'vehicle', a 'vehicle' is something used for 'transporting' stuff, etc.), ultimately, symbols need to be related directly to an outside reality (what does 'something' and 'stuff' actually mean?). In the same way that drilling down into a Wikipedia page by clicking on its Wikipedia links tends to draw you inexorably into the deep waters of Wikipedia's Ur-page: 'Philosophy', or that trying to deal with a youngster's stream of 'why?' questions can suck you into knotty existential conundrums ('But Mum, why does energy exist?'), and trying to bottom out the meaning of something by appealing to other things that have meanings turns out to be something of a fool's errand. All of the rules and symbols in Searle's rule book might capture the *syntax* of Chinese, but not its *semantics*.

For me, one attractive way of approaching this problem has been to contrast the kind of things in the world that don't have meaning with the kind of things that do. Unlike a sentence, the sound of some waves crashing on the shore doesn't mean anything. The wave sounds are not about something, but a sentence is. A sentence can be true or false, can be well-formed or malformed, can work or fail to work, but wave sounds just *are*. Sentences have jobs to do but waves do not. So where do jobs come from? One answer, most brilliantly expounded by philosopher Professor Ruth Millikan, is: biology.

Consider: Your heart has work to do. To keep you alive by pumping blood around your body. There are lots of things that your heart does, but pumping blood is its job. Your heart makes a bumpety-bump noise, but that is not really its job. Your heart is red, but that is not really its job either. You can tell this, because you could imagine being born with a green heart or a heart that made a squeak and, so long as it still pumped blood just as well as a regular heart, you would not have grounds for saying that your heart was malfunctioning. The reason that pumping blood is your heart's job is that it was by pumping blood that your ancestors' hearts contributed to their having offspring and, ultimately, to your own existence. Pumping blood is the evolutionary function of your heart because that's what it was selected (by Darwinian evolution) to do. Hearts were not selected to be red, or make a bumpety-bump noise. Those aspects were only incidental to the reproductive success of your parents and their parents before them, and so on. And, consequently, now we have hearts that can, unlike rocks or wave sounds, 'go wrong', fail to do their jobs, be malformed. When a heart breaks, it must be fixed, but waves break all the time without issue.

While Searle appealed somewhat mysteriously to some unspecified 'causal properties' of the brain as the thing that separates unthinking machines from thinking people, Millikan provides a much more complete story. She argues that words and sentences are like hearts and lungs, they have a function to perform that was established by their history, by the selective pressures that have shaped the (brain) mechanisms that we use to produce them. This special kind of causal grounding, causal connectedness, and causal history is what enables meaning. A human child raised in China might come to know and understand a Chinese language through a combination of its evolutionary history and its normal personal experience in a language using community. The man in the Chinese Room doesn't have this combination and, as a result, it's not surprising that understanding and meaning are missing.

In taking this tack, Millikan is *externalising* meaning: the meaning of a symbol is not intrinsic to it, not to be found in its shape or its colour (my signature is my signature even if it comes out looking more like yours), or even in the shape and colour of other symbols alongside it, but is rather to be found in the history of external events in the wider world that gave rise to it. These are things that cannot be determined just by looking at the symbol, or even inside the symbol maker's head, or its memory banks. Consequently, by exporting considerations of meaning into the wider world in this way we lose our ability to intuit clearly about the *locus* of understanding. Where does it lie? Is it in the little neurons in a man's head, the whole of the little man, the little man plus his book of rules, the man and the book and the desk, or the little man and his house and the whole world?

I once sat in an audience with Professor Maggie Boden, historian and philosopher of cognitive science, listening to an invited seminar by a very dynamic neuroscience professor. After presenting studies on the function of different parts of the human brain, he finished with a rhetorical flourish, declaring that since the different jobs of different parts of the brain were almost worked out, all that remained was to find the part of the brain that co-ordinated and gave meaning to their activity. Exasperated, Maggie raised her hand and asked, mischievously, whether he had perhaps thought of investigating the pineal gland. Ignoring some scattered chuckles, the speaker dutifully noted it down as a possibility, obstinately missing the sardonic reference to Descartes, who believed the pineal gland to link our material bodies with our immortal souls. But we should perhaps sympathise with the neuroscientist and the AI engineer, since fully embracing the "extended mind" is sometimes rather like reaching the end of a sci-fi novel where 200 pages of firm technologic plot dissolves into a brief befuddling swirl of opaque and fulsome prose, warping space and time as it purports to pierce the mind of God.

Annie Clarkson's 'The Rooms' takes us on a very different journey and, like Millikan's, it is one that meshes meaning with work and with biology. Her protagonist, like many of us, is looking for a job that rewards us with some authentic meaning as well as remuneration. But Hannah's job is phoney, and the girls that Hannah encounters in the The Rooms are fakes, working jobs that have been carefully and deliberately designed to be meaningful only for their customers.

It may be true that no-one thinks an algorithm will really burn the neighbourhood down, but the notion that following rules and regulations at work can ensure that a real job gets done is uncontroversial: 'A job is a job'. Charles Babbage, the designer of the very first automatic computer, was taken with the idea that a factory could be thought of as a huge machine, with people and computers and lathes and presses all simply its mechanical cogs, getting the job done by going through their motions. But the artisans capable of machining components fine enough for his Analytical Engine design were not minded to play their part in this scheme and the computer went unbuilt. More recently, attempts to incentivise and regulate the caring professions have raised the question of whether formal procedures, metrics, tests and rules can have the 'causal properties' of ensuring that we, our parents, and our children will actually and really be protected, be cared for, and be taught, rather than merely processed. Famously, the oldest profession is also a caring profession, and the tension across the gap between real intimacy and merely 'going through the motions' is surely at its heart.

Like brains, lungs and (according to Millikan) sentences, sex also has its proper functions. Sex is for fun and, ultimately, for reproduction. But these links are deliberately broken inside the Chinese Rooms. Can they be mended? Do Myla, Alina and their co-workers even have reproductive systems behind their sex organs? Do they, for instance, menstruate? Could the 'moon cycle' that Hannah and Samuel discuss, come to have meaning for them? Has it already? The story culminates with Hannah choosing a route forward. Will she succeed? Can we make meaning for ourselves? Or are we subject to the sense and senselessness imposed by the external world. No matter how closely Myla and Alina learn to mimic Hannah's behaviour, they are never expected to actually begin to experience the emotions that drive it. That wasn't part of their program. But could learning, like evolution, ground behaviour and confer meaning on what was initially meaningless? Just as some habitual cycles engender a sense-making autopoesis, perhaps new kinds of meanings can be co-produced in Chinese Rooms.

If He Wakes

Margaret Wilkinson

HER FATHER'S ROOM IS cold, then hot. From his window, overlooking a suburban garden, he surmises he is not in town. White clouds lie motionless in the sky. *How did he end up here*, he thinks, *and why?* Although he has trouble walking, he decides to leave. He wants his hat and his coat. Anna imagines every detail. He's going to catch a train into town. Hail a cab. He has the nagging feeling that there's somewhere he's meant to be; someone he's meant to meet. In town is the apartment he left, full of his belongings, his souvenirs that represent a lifetime of collecting. Her father needs these things, Anna thinks, his old copies of *Scientific American*, for example. And his silver plated claw cutters in case he is served lobster for lunch. After lunch, which is meatloaf, not lobster, her father has a sleep. When he wakes (if he wakes) he will stare out the window watching the sunlight descend the aluminium sided houses opposite, whose residents are less than pleased to live next door to a nursing home.

The shadows lengthen, but it still may not be too late. Not if he leaves now. She imagines him crouching on the window sill. He is on the first floor and it is an easy drop to the lawn below. He creeps across the well-tended grass and climbs a suburban hedge. Soon he's on a dark avenue bordered on either side by very tall, closely-planted fir trees, slipping on the needles in his haste. And suddenly he sees, without riding the train, or grabbing a cab, the beautiful shining asphalt city he so thoughtlessly left – the post office steps he falls down when he is no longer coping, his old office and waiting room, the restaurants he loved. The Greek diner with the wobbly tables is, of course, gone; the bakery, the dry cleaners. Anna knows this. The synagogue is now a bank; the bank with its lead-lined security boxes, a chicken restaurant where seating downstairs in the vaults makes conversation impossible.

Anna is six. Her sister is five. Their father, in his prime, gathers together two small saucers, a basin, some sawdust and a jug of water. They are going to do an experiment. He places the two saucers on the table and carefully fills them to the brim, one with sawdust, the other with water. (This is how we understand the world around us. This is how the world behaves.) He asks Anna to add more sawdust to the sawdust saucer. She does so and the sawdust stands up in a heap. Then he asks her sister to add more water to the water saucer. Water will not stand up in a heap. Some of it runs away over the edge of the saucer and drips onto the table, and down to the fitted carpet. (Because water flows, we call it a liquid.) He asks Anna's sister to try and pick some water up in her hands. More water spills onto the table, the carpet. (We cannot pick water up as we pick up other things.) Then he pours the water from the saucer to the basin. (We can pour water from one vessel to another because it flows.)

'Which way does water flow?' He looks at his two daughters. Anna shrugs her shoulders.

'Down,' her sister calls out. 'It flows down.'

When Anna arrives, he isn't in his room. Alone with his things, she wonders what she would take as a hypothetical souvenir. A tea cup and saucer so recently trembling in his hands? His pillow? Pillow case? The leftovers from his lunch tray? A scrap of paper upon which the name 'Didi,' is written in a jittery hand?

She waits for him and when he doesn't come, she opens the door and looks down the corridor. And there he is, slowly advancing. She recognises his much admired nose, his big sorrowful eyes, his disappointed mouth; the past written on his face as it will be written on hers. His skin is thin and papery as if he's been mummified in the cool, dry atmosphere of the chicken vault which used to be a bank. The moment he sees her, his features droop a bit more and she wonders, is this real or fatuous posing? You never know with him. Catching sight of her, he walks more slowly, more painfully, gripping a wall-mounted handrail tightly, one foot raised in its ugly black boot with Velcro fastening. 'How are you?' she asks. He doesn't seem to recognise her, but takes her hand and holds it firmly. *He's play-acting*, she thinks. *He must be. His hands are cold and she is, for a moment, trapped in his icy grip.* He searches for her name.

'You sound like your mother,' he says finally. 'Aren't you ashamed?' He bangs his stick on the floor. He looks around. There is somewhere he's meant to be, he thinks, someone he's meant to meet.

'I hear you're not eating,' she tells him. She opens her coat feeling very warm, despite his coldness, waiting for whatever will come next. She tries to time her visits so as not to run into her sister. She is not sure what the nurses in their butterfly uniforms, or the drab-suited carers, have been told. She is fearful of their condemnation and worries that they talk about her behind her back.

'She didn't need to know,' they all whisper when she passes. Everyone blames her.

As soon as she hears the words 'thought' and 'experiment,' Anna tunes out. She would rather be riding her bike. 'Can a particle exists in all states at once?' her father asks as she edges to the door. She is ten years old. Her sister is nine. It is a beautiful day. Bars of yellow light lay over the fitted carpet. Her sister steps on one.

Her father is saying the name, 'Schrödinger.' Anna is putting on her coat. Then a cat is mentioned, the consolation of ordinariness, fur and whiskers – and she hesitates. A cat, according to her father, is put into a lead-lined box along with a radioactive sample, a geiger counter, a hammer and a bottle of poison. Is she paying attention? He goes up to her and takes her by the shoulders. She looks at the carpet. The amount of poison needed is carefully calibrated to the weight of the cat. 'If the cat weighs x pounds and it takes y number of drops of liquid cyanide per pound to kill it, how many drops of cyanide does Schröndinger need?'

Anna shrugs her shoulders.

'I know, I know,' her sister says.

'In one hour, which is the amount of time the cat must remain in the box, the radioactive sample has a 50% chance of decaying,' her father informs them. They don't have a cat. Is he trying to tell them they are getting a cat? Anna would prefer a dog, but wants to please him. *If he wants a cat*, she thinks, *let him have a cat, a cat would be nice*. Anna feels a mixture of interest and disinterest. Should she stay to hear the rest of the story? Or not? Her mother's knives clatter in the kitchen. Then there is the smell of frying onions.

'She didn't need to know,' they will whisper.

Her father is a man of science. He towers above her. For years she thinks, or has been told, that he is an ophthalmologist, a medical doctor who can diagnose and treat, perform surgery, and cure eye disease; then she learns he's not an ophthalmologist, he's an optometrist, who is after all a healthcare professional, sight-testing, prescribing and dispensing corrective lenses; finally she hears he's not an ophthalmologist, or an optometrist, he is an optician, a mere technician who sells and fits lenses and frames in a shop – a shop with a waiting room, he claims.

'If the geiger counter detects that the radioactive sample has decayed, it triggers the hammer, which smashes the bottle, releasing the poison and killing the cat.' Although this is upsetting, there's a cartoon simplicity to the experiment Anna likes. She thinks hard about what her father is saying and tries to understand droplets of cyanide slowly or quickly falling. He speaks in the reverent tone he reserves for science – and all the while, while he is talking about cats and boxes, geiger counters and cyanide, he is on the way to his future. And she to hers. She's already thinking about something else. What souvenir, for example, she would take from Schrödinger's experiment. The hammer?

'Only when we open the box will we know if the cat is alive or dead.' The sentences continue to roll out of her father's mouth. 'Because at that moment the superposition of the cat collapses.' He looks hopefully at his two daughters. 'Who remembers what superposition means?'

Anna shrugs.

'I don't have all day,' he tells her. There is somewhere he is meant to be, someone he is meant to meet.

'I know,' her sister says.

Anna doesn't want to have anything more to do with this nasty experiment. Now he's talking about wave particles, which is boring. 'These particles do not obey Newton's laws,' he says. 'The rules we use to explain the ordinary world cannot be used to explain electrons or atoms. Who remembers what a wave function is? Anna?'

'I know,' her sister says. 'A wave function shows all the positions an atomic particle can be in. It can be in all positions at once, which we call superposition. We don't know which until we measure it.'

'Good girl,' her father beams, leaning forward and giving her sister a hug. 'Now, who can tell me about the Copenhagen Interpretation?'

Anna's father knows Schrödinger personally, or feels as if he knows him, and refers to him as 'Erwin.' (He is impressed by Erwin's unconventional morality, living as he does in Oxford and later in Dublin, with both his wife and mistress, even fathering daughters from different women.)

From Schrödinger's lab Anna's father takes a souvenir, a test tube commemorating their alleged first meeting, which he carefully hands around at Christmas. Before his death in 1961, Anna's father claims to meet Schrödinger again in Vienna. It is at this meeting that her father catches tuberculosis from Erwin, which ends his hopes of being an ophthalmologist forever, but he isn't bitter. Before Anna is born, half her father's right lung is removed to a jar he keeps in the hall closet behind the hats and winter scarves, where it floats, pink and speckled grey, in a disinfectant solution. (She is frightened of this closet for her whole childhood and never hangs up her coat, irritating her mother.)

It was while in a sanatorium that Schrödinger formulated his famous wave equation crucial to the understanding of the behaviour of subatomic particles and light. (Anna's father, being a trained optical worker, could tell Erwin a thing or two about light.)

There is a painting in their front room, a dark scene of men around a campfire, firelight inexpertly illuminating their faces. Schrödinger painted that, according to her father. Many years later she reads a quote concerning quantum mechanics. 'I don't like it,' Erwin Schrödinger says, 'and I'm sorry I had anything to do with it.'

Care assistants change his food and water, warning each other in their own language, which is not his language, about him, or so she imagines. They walk on either noisy or soft shoes, discourage weeping in daylight and turn a deaf ear at night. Sometimes they remove a limp body from her father's corridor, or elsewhere.

In the mirror his reflection trembles. He changes the expression on his face from neutral, to thoughtful, to pitiful, this problem-solver, this amateur, this science buff. He can (could) solve all sorts of number problems, but he cannot solve the problem of his life. At night the back garden of the nursing home, called a yard, is floodlit. Something soft and heavy falls to the ground from the window above his own. Another resident escaping.

Alone in his room he senses a door creak open, then closed. He turns slowly around. There's a strange antiseptic smell. Are they coming to sedate him? There is a woman, even here, a resident, who has peaked his interest. One afternoon she is sitting in the dayroom which is crowded with armchairs. He enters, makes a show of looking around for a seat, choses a chair beside hers. A moment later he's told her he's a retired ophthalmologist and is helping her with her eye drops. Then he takes her hand. 'May I?' He still has a headful of hair which some women find attractive. No paunch. Anna watches from a visitor's chair. She's come to tell him she is leaving the country, but he's ignoring her. He looks instead at the woman beside him, who's called Edna.

Edna wears a dress with scalloped shoulders. She has terrible blue-black hair, age-freckled hands, silver nails. He asks her about her favourite foods, which is a hot topic of conversation among the oldsters disappointed with their current meals.

Her father sends back cold eggs, underdone toast, as if he were in a hotel restaurant. His favourite foods: peanuts in the shell, salami, fried scampi, blue cheese, lobster. 'Do you like lobster?' he asks the stooped little old lady with blue-black hair who is sitting beside him. 'How about science? Do you like science?'

Edna doesn't know about science or lobster, but says she likes popular music. He says he likes popular music (which is news to Anna who has never known her father to like popular music). When cups of tea are served, he's awkward, charming, helpless. Edna steadies his cup for him. He asks if she likes cats. She does. Meanwhile, across the lounge, a woman her father recently took up with then abandoned, glares at them. Carelessly, he lets his paper napkin slip from his knees. He might be wondering if Edna has a radio in her room. He knows how to build radios and can explain how radios works. Edna listens to him patiently, her eyes shining. He will ask, when they are alone in her room, to examine her eyes. He is concerned about the mist she says she has noticed settling over all the electric lights in the home, the overheads and table lamps, which might be the onset of macular degeneration, or dust. It is unlikely they'll be allowed walking privileges, but if they are he'll take Edna to the sea at sunrise and sit beside her on a bench, soothed and spellbound. He is half in love with her already. His daughter tries again to say goodbye. She is relocating to a faraway place where the wind grabs people's hats from their heads, where she will have to imagine her father in her life. But he is preoccupied. Whatever is unresolved between them will stay unresolved.

'I might be old,' her father says, 'but I'm not dead.' Or is he at this moment floating down a corridor, his arms outstretched, frightening those care assistants (more than you might think) who have second sight and can apprehend his ghostly form walking through walls, plate glass windows, doors? Now she sees him lying in bed with a peaceful expression, dying as we would all like to die – a butterfly sprouting from his lips; his eyes big and sorrowful one moment; empty the next.

When Anna was a girl, her father wished to be buried at sea. From an old sea chantey he liked listening to, he got the idea of having himself wrapped in a jacket made from a material used to protect exposed objects. 'Wrap me up in my tarpaulin jacket and say a poor duffer's laid low. Send for six salty seamen to carry me, with steps mournful solemn and slow,' he sang.

'Where'll we get six salty seamen?' Anna and her sister used to joke, when they still joked, when they still talked to each other.

For all Anna knows he is already dead – and no one's told her. He's in a new nursing home now, his fourth or fifth, in a country she left many years ago. Anna cannot contact him because her sister will not reveal where he is. Nor does he have a mobile phone. (If he had a mobile phone, he would be ringing emergency services, threatening to kill himself. He's done it before.) Neither her sister, nor any of her other relatives, want her to contact her father. They close ranks against her.

Anna has not spoken to her father in two years. She goes through many e-mails and finds the name of the last nursing home he resides in, discovers the telephone number online, and decides to ring. But she doesn't ring. Every day she finds another excuse not to ring. She calculates the time difference. It's too early. Too late. Until she rings, she does not know whether her father is alive or dead. She phones the home. He's alive. She phones the home. They are very sorry, he's passed. Can he be both alive and dead? As long as she postpones calling he is. She might be afraid to learn that he is indeed dead. Or she might be afraid to learn that he's still alive, stewing, seething with neglect, pretending not to know, or remember her, when he hears her voice.

'Dear Dad, How are you?' She writes a letter he will never answer. Or maybe he will answer in his jittering scrawl. He will say he cannot help himself and he dreads to think what the end will be, but wants to stop hiding, stop telling lies, stop leading a double life. 'Are we going away together or not?' he'll ask, thinking she is someone else.

In her father's late middle age, he takes Anna, for a birthday treat, to one of the elegant seafood restaurants he likes, where there's candlelight and the waiters all wear short red coats. One comes forward and greets them. He is a close-shaven man with a puffy face whose hands will shake when he serves dinner. Her father pats him on the back and calls him Dominic. It is Anna's eighteenth birthday. Dominic wishes her many happy returns of the day. As they are lead to their table, Anna wonders if the other diners will think she is his date, her handsome, clever, well-dressed father. Never. Maybe.

The table is laid with silver and the napkins are linen. When they order lobster, elegant little forks like silver picks, lobster scissors, claw crackers and tiny bowls of cut lemons and melted butter appear. Her father wears a smart and expensive suit. (How can an optician afford such a suit?) The waiter, Dominic, solemnly ties a paper lobster bib around his neck.

'Mademoiselle?' He unfurls her bib, his too-sweet breath lifting strands of hair on the back of her neck as he bends close to tie it. There are fishing nets arranged tastefully on the walls and French curtains the colour of sea water at the windows. Her father examines his claw cutters and snaps them experimentally in the air. Then he begins cross-examining her about the biology class she is taking.

Halfway through the meal, a lovely young woman brushes past their table, turns, stops, her eyes bright as jewels. 'Look who's here. Look,' she cries. Or is it him, her father, who cries out to the young woman? Anna remembers every detail of her eighteenth birthday dinner, but not this one.

The woman takes her father's hand, lightly kisses his cheek – a butterfly kiss on the cheek, nothing more. He stares at her abstractly as if trying to remember her name. A current patient? He invites her to join them for a drink. He clicks his fingers in the air to summon the waiter and orders the woman (although she protests) sizzling prawns which are brought to the table in a frying pan along with a Beefeater Gibson.

Anna, whose birthday it is, shrinks back and looks around fearfully, as if she were the intruder, the uninvited guest. There are little rolls on the table. Her father slits one open with the point of his knife, butters it and begins to eat slowly, speaking to both his daughter and the young woman, who he calls Didi, in a tone of light irony.

Didi turns sweetly pink, her glamourous little eyeglasses (fitted by Anna's father) steaming slightly in the moist atmosphere of sizzling prawns. The whole arousing, disturbing sense of her, this Didi, alarms Anna, while her father, unaware, explains the difference between shrimps and prawns.

Anna has already taken a souvenir, a lobster fork, and wishes she hadn't, but it is impossible to put it back. Without saying a word, she gets up and makes her way unsteadily between the tables to a ladies toilet where she is sick.

When Anna returns to her seat, she sees that Didi's eyes are filled with tears. 'I don't understand,' her father is saying. 'What is it you want?' When they see Anna, they both begin laughing. 'We're just discussing dessert,' her father explains. Didi doesn't stay long after that, but the meal is ruined. Her glamourous cats' eye glasses need readjusting. Anna's father slips his diary from a breast pocket.' When would you like your next appointment?' he asks. Then he walks Didi out of the restaurant, and gets her a cab, leaving his daughter alone at the table.

Anna doesn't tell her mother straightaway. She waits a year. Two years. Three. At night she lays awake thinking. If not her, who will blow the whistle? While she waits she becomes convinced this is not the first time, nor will it be the last. Not for one moment does she think she's got it wrong. Her mother might even suspect and would be glad to know. Her father cannot be both a husband and another woman's lover.

In the end, Anna decides to tell. She breaks her mother's heart, according to her sister. After which her mother complains of sleeping badly, her broken heart beating uneasily. 'Please,' she says. She cannot say more because she is crying.

No one knows where she got it. You need a license to buy it. But it wouldn't have cost much. Anna's mother is not a big woman. In human beings a fatal dose can be as low as 1.5mgs per kilogram of body weight. (Liquid cyanide, decanted into a thin-walled glass ampoule, and covered in brown rubber to prevent breakage, could be carried in the mouth, shaped like a false tooth. Or hidden in the filter of a cigarette as depicted in spy films and novels.)

Anna's mother looks surprisingly old when she is found, old and forlorn, her chin sunk on her chest.

Recently physicists' have a new theory, her father would be interested to know. Or perhaps he already knows. When

Schrödinger's box is opened, the observer and the cat split into two realities. In one, the observer sees a dead cat; in the other, an angry cat.

She decides to phone the nursing home. 'Hello, I would like to enquire about my father,' she practices saying, clearly and calmly, so that when the phone is actually answered by a nurse, or care assistant, in a different time zone, many thousands of miles away, she will not panic.

She dials, or rather punches in the numbers. The phone rings its foreign ring, then it is picked up.

'Hello,' she says. 'Hello.'

Afterword:

Schrödinger's Cat

Prof. Steven French
University of Leeds

SCHRÖDINGER'S CAT IS ONE of the most famous thought experiments in all of science. Physicists, in particular, like to use these imaginative exercises in various ways and for diverse purposes. Einstein, for example, with his 'Chasing a Beam of Light' and 'Free-Falling in a Lift' thought experiments (see pp.241-253 and 197-210), was a master of using the technique as a motivational or heuristic tool, that is to say as a way of getting to the conceptual heart of some core issue in physics and thereby helping to generate an entirely new theoretical framework.

Schrödinger's aim with his famous cat-in-a-box thought experiment was different (and perhaps it's for this reason that it sometimes called a 'paradox') – he wanted to highlight what he regarded as a 'ridiculous' (his word) feature of that understanding of the new quantum physics that was coalescing around the work of Born, Dirac, Heisenberg and, most importantly, Niels Bohr. This understanding is now typically referred to as the 'Copenhagen Interpretation' (thanks to the location of Bohr's institute), although that name came much later and the interpretation should by no means be regarded as monolithic, composed as it is of different strands and drawing on different philosophical views. At its core is the idea that the state of a physical system is completely described by something called a 'wave function'. If the system can be in more than one possible state of a given property, then the wave function will represent a superposition of those states. So, for example, if we consider the property of the spin of a particle, it turns out that the possible states are spin 'up' and spin 'down' (imagine a little arrow pointing along the axis of spin – although the particle isn't really 'spinning' as we would usually think of it). Then the wave function for the state of spin of the particle is a superposition of spin 'up' and 'spin 'down'. And likewise for all properties of all physical systems.

It is this aspect of the new quantum physics – the idea that systems could be 'in' these superpositions – that so bothered Schrödinger (even though he played a big part in the development of the theory); it also greatly bothered his friend Einstein. Indeed, it was Einstein, with his facility for thought experiments, who suggested a forerunner of the cat experiment in a letter to Schrödinger[22] in which he imagined a keg of gunpowder existing in a superposition of 'exploded-and-not-exploded'. But it was Schrödinger who captured everyone's imagination with his cat, placed in a box in which there is also a sample of radioactive material, a Geiger counter, and a flask of hydrocyanic acid. Atoms in the radioactive material have a certain probability of decay; if one does, it will trigger the Geiger counter which causes a hammer to break the flask and the cat dies. That radioactive decay is explained by quantum physics (via something called quantum tunnelling) and the wave function gives the probability that it will occur. But the crucial point is that, according to the above understanding of Bohr and others, the state of the atom and hence of the atom + Geiger counter and hence of atom + Geiger counter + hammer + flask + cat, that is, of the whole system, must be described as a superposition of atom-not-decayed-Geiger-counter-not-triggered-flask-not-broken-cat-alive and atom-decayed-Geiger-counter-triggered-flask-broken-cat-(sadly)-dead.

And this is what really bothered Schrödinger: are we prepared to say that inside that box, the cat is neither alive nor dead, but in this weird superposition of both states?! It seems utterly bizarre that reality could be that way yet that is what quantum physics, on this 'Copenhagen' understanding, seems to tell us.

And the reason it seems bizarre, of course, is that we don't seem to observe such superpositions in practice. When I look at the small dog snuggled up in the armchair next to me, I don't see it as in a superposition of states. Nor do I see any such superposition when I look at insects, microbes, even large polymers and the like through a high-powered microscope. Likewise, if we were to run such a horrible experiment and then open the box, we would expect to find the cat either (furiously) alive or, sadly, dead. How can we align what we observe with what the theory tells us? (This is the 'Measurement Problem', beloved by philosophers of physics.)

One of the earliest answers placed the emphasis on us, as conscious, sentient observers:[23] it is when we make an observation that the superposition collapses and we obtain a definite result. So, it is when we open the box that the superposition of cat + everything else collapses into one or other of the two alternatives and we find either a dead cat or one that is very much alive. This answer then generated a whole slew of pop-science books and articles about the subjective nature of reality, the role of consciousness in determining what there is, the relationship between quantum physics and Zen Buddhism and so forth.

But it was eventually knocked on the head by certain philosophers, among others, who questioned the mental-physical dualism that it presupposes and asked how it is that the mind could effect a collapse of a quantum superposition? (As it turns out, not all of the physicists who suggested this answer were as philosophically naïve as these philosophers seemed to think. But that's a story for another time!)[24] And more pointedly, perhaps, if we accept that quantum physics applies to every physical system, including the universe itself, then what causes the wave function of *the entire universe* to collapse from a vast superposition into the definite features that we observe, as highlighted so wonderfully and beautifully by the images from the Hubble telescope?

Reflecting on this question led the American physicist Hugh Everett III to propose that there was no such collapse at all – that what happens upon measurement, or indeed any physical interaction, is that reality 'branches' into the various alternatives.[25] According to this view, when we open the box we do indeed see either a live or dead cat, but not because of any collapse of the wave function, whether under the impact of a conscious mind or anything else, but rather because before we even did so, reality split into two branches: in one the atom has not decayed, the vial is not broken and the cat is alive and in the other, the decay has occurred, the poison has been released and the cat is dead. Understanding these branches as worlds yields the Many Worlds Interpretation of quantum physics, where 'many' = a *huge* number, but unlike in the movies or science fiction novels, the physics itself tells us that we can't hop from one such world to another!

This interpretation has also generated considerable controversy, not just because of the idea that reality is constantly fracturing all around us, but also for more technical reasons: quantum physics yields the probabilities of the outcomes of observations (50-50 in the case of the cat). But what sense can we make of such probabilities when all outcomes are realised, albeit in different worlds? Efforts to answer this question have led researchers into the arcane realm of Decision Theory in an attempt to articulate what probability can mean in this context and at this point the sharpness of the cutting edge can really be felt![26]

Literature offers thought experiments too, of course, and Margaret Wilkinson's story invites us to reflect on multiple uncertainties at different levels. To what extent does our behaviour change when we are being observed? The little dog in the armchair next to me had an operation on its leg – long since healed but when it sees me watching, it suddenly starts hobbling on three legs, in 'poor me' mode! Likewise, but more seriously, Anna's father's posture changes, his features droop a bit more, when he catches her observing him in the care home.

And how much do we know about our parents, our family, our friends? Anna thinks, or has been told that her father is an opthamologist; then she hears he's an optometrist; and finally, that's he's a mere optician. Her uncertainty collapses down through various alternatives, to what seems like a mundane outcome. Yet it is one that offers the opportunity to meet the great Schrödinger and even to catch tuberculosis from him somehow – how could this have happened? Considerable uncertainty remains.

But more profoundly, it is Anna who is the radioactive atom in this tale, waiting years to decay, before telling her mother of her father's infidelity (although even that is left unclear) and causing her mother to take the poison and for the family to collapse. Nevertheless, the state they are in towards the end of the story is unresolved and un-collapsed in other ways. Not least, the question of whether her father is alive or dead after all this time. When she calls the nursing home and the phone is picked up, does her conscious intervention collapse the superposition of possibilities or does reality split into branches, each with its own definite outcome…?

Of course these are just metaphors and we should be wary about attaching too much significance to them. Not all uncertainty is quantum in origin – sometimes it is just down to our ignorance. Even if quantum theory applies to all physical systems, as most physicists accept, recent research on something called 'decoherence' shows how when quantum systems (such as the particle produced by a radioactive atom decaying) interact with their environment, the quantum nature of the system's state may be effectively suppressed.[27] This suppression is typically very strong when it comes to macroscopic systems, which helps explain why we don't observe it in such cases. Some advocates have maintained that in itself decoherence resolves the issue presented by Schrödinger's Cat but most commentators now agree that it needs to be combined with something like the many-worlds view, where it helps explain why we observe the 'worlds' that we do, namely ones in which macroscopic systems have definite locations, for example, and cats are either alive or dead. So, Anna's father, as a huge, macroscopic system, is already alive or dead when she calls and her picking up the phone is not going to change that.

And going the other way, when faced with the other-worldly strangeness of quantum physics, we naturally clutch at whatever comes to hand, whether metaphor or metaphysics, in an attempt to get our heads round it. Schrödinger, again, coined the term 'entanglement' to describe the peculiar quantum correlations between two particles that interact, referring to it as *the* characteristic trait of quantum theory. But how to understand this strange entangling of systems remains a deep problem, to which a whole range of philosophical and physical tools and devices have been applied. His cat thought experiment allows quantum theory to intrude into the everyday and invites us to reflect on its strangeness and how it goes beyond any current metaphor or metaphysics.

Notes

22. P. Halpern. *Einstein's Dice and Schrödinger's Cat*, (New York: Basic Books, 2015,) p.140.

23. John von Neumann, *Mathematical Foundations of Quantum Mechanics*, (Princeton, Princeton University Press 1932, repr. 1955), first published in German in 1932 as *Mathematische Grundlagen der Quantenmechank*, (Berlin: Springer); Edward Bauer and Fritz London (1939), *La Théorie de L'Observation en Mécanique Quantique* (Paris: Hermann, 1939), rep. in J.A. Wheeler and W.H. Zurek (eds.), *Quantum Theory and Measurement*, (Princeton: Princeton University Press, 2014).

24. The problem with mind-body dualism (the belief that mental states constitute something *more* than merely physical states: e.g. they correspond to a different *substance* or to different *properties* than the physical substances or properties of the brain) is that it remains unclear how mental substance interacts with

physical substance or mental properties with the physical properties of the brain. And this problem is highlighted in the case of the cat: how does consciousness cause the cat's wavefunction to collapse into one state or the other, alive or dead? However, one of the proponents of this approach was Fritz London, a brilliant physicist who died, tragically, too early to win the Nobel Prize (see K Gavroglu, *Fritz London: A Scientific Biography.* (Cambridge: CUP, 1995.)). Before he became a physicist he studied as a philosopher, most notably, in the tradition known as *phenomenology*. Very briefly, this can be framed as 'an enquiry into the essential structures of consciousness' but what is important for our purposes is that it can be understood as eschewing the kind of mind-body dualism in focus here. As far as London was concerned, consciousness does not intervene in some mysterious way to collapse the wave function; rather consciousness itself must be included in the wavefunction but then it has the capacity to separate itself and, as he put it, 'set up a new objectivity', which amounts to the cat being either alive or dead (for further details see S. French 'A Phenomenological Approach to the Measurement Problem: Husserl and the Foundations of Quantum Mechanics', *Studies in History and Philosophy of Modern Physics* 22 (2002), pp.467-491. This is the understanding that lies hidden in London's famous piece with Bauer and which has been overlooked by many commentators on this issue.

25. Hugh Everett, 'Relative State Formulation of Quantum Mechanics', *Reviews of Modern Physics* 29 (1957), pp.454-462.

26. Decision Theory is the theory of how we make, or should make, decisions, particularly in situations of uncertainty. Typically, it requires you to assess all the possible outcomes of a decision, assign a certain value to each one and a probability that this outcome will be realised, then combine the two to get the 'expected value' of that outcome. On one understanding of probability, it has to do with what rationally compels your degree of belief about a particular outcome (e.g. in the throw of a die, it is rational to believe that the chance of getting a 3 is 1/6). In the case of the many worlds interpretation, this understanding has been taken up to show that you are rationally compelled to set

your belief in a particular outcome (cat dead, say) to the 'weight' of the branch or world containing that outcome (D Wallace, *The Emergent Multiverse* (Oxford: Oxford University Press, 2012)).

27. See, for example, M Schlosshauer, 'Decoherence, the measurement problem, and interpretations of quantum mechanics'. *Reviews of Modern Physics* 76: 4 (2005), pp.1267-1305.

People Watching

Claire Dean

THE YOUNG COUPLE AT the bar looked happy enough – she played with the zip on his jacket. His hand cupped her elbow. But I could hear what she was saying, and his gaze was somewhere above her head. 'If you'd just told me about the flask in the first place, it wouldn't have been a problem.'

The man behind the bar wore a grey wool hat pulled right down to the rim of his glasses. After every customer, he took time to smooth the pages of his paper back out across the bar. He ducked down behind the rack of paired, plastic-wrapped biscuits to read.

At the table across from mine a middle-aged couple struggled with three young children. The baby's cry sucked air in around it. The mother threw a bottle across the table at the father. I waited for her to throw the baby as well. A heap of coats, soft toys and changing-paraphernalia had created a bunker around the family, which two little girls destroyed with delight again and again. They wore fairy wings over their aran jumpers and bounced about in bright pink wellies seeing who could fly highest. Their screeching didn't needle me as much as the forced politeness in their parents' admonishments. 'Chloe, sweetheart, please don't scratch your sister's face.'

It was the elderly couple I was drawn to most. The man had his back to me, but I could still get a sense of him; see how he was with her. The contour of his back and arms suggested he was holding her even from across the table. Her face was still; its wrinkles suggested past smiles. I tried to imagine looking at someone with such trust. He poured her tea a little at a time into a second cup and helped keep it steady as they raised it together to her lips. They didn't have need for words.

I slipped out of my booth and crossed to the bar. I'd been noting what others had ordered and decided to treat myself to the day's special, hot chocolate with rum. The spirit bottles hung neck downwards. Their contents made me think of the different shades of preserving fluid that hold tiny corpses in jars. The bartender removed a mug from beneath the rum bottle to pour my measure and then swiftly replaced it. The little plastic tap on the neck was dripping rum with such regularity you could set a watch by it. How long it would take for the bottle to empty that way, I wondered, and would the man pour the rum back into the bottle, or drink it himself?

Two crumpled fairies, mouths smeared with chocolate, jumped up beside me to peer into the fish tank next to the bar. 'Look, look, look!' The tank's inhabitants were brown and lethargic. They worked their way round the edges of the tank sucking at green scum. A faded sticker on the glass said 'Visit Lakeside Aquarium Today. A Fantastic Fun Family Day Out. Open 9-6, 7 days a week'.

'Chloe, Lucy. Sit. Down. Now.'

The girls pinballed between the tables. A ripe smell suggested the baby needed changing. The mother dispensed mat and equipment and laid the baby on the table just as the girls piled back in around her.

The young man seemed compelled to watch. He looked confused. His girlfriend didn't glance up once from touching the screen of her phone.

For a moment, I saw them all pinned like specimens to the narrow tables. The two girls' wings fluttering as if they could not be held down. I would preserve the elderly couple beside each other, lying in companionable sleep as they must have done for so many nights of their life.

I wondered what they all thought of me sitting there alone. Stretching out sips of my rum-laced hot chocolate. Would they see how much I struggled with it: not wanting to drink it too fast because then the warmth would be gone, but knowing in my hands it was losing heat anyway. Or perhaps they hadn't seen me. I often suspect aloneness renders one invisible.

Condensation had claimed the windows, jewelling the dusk outside. The elderly man stood. He pushed the cups to one side of their table and started to ready his wife's things. I watched him put on her grey wool coat, a pair of pink mittens, a white scarf. I thought of a father dressing his little girl. Buttoning her up against the world. They both moved with arthritic tenacity. I marvelled as she stood at the curve of her back. She was curling into herself.

The last mouthful of hot chocolate was bitter and grainy. I watched the children flit about for a few more minutes before pulling on my coat and heading outside.

The air was so cold it hurt. At the top of the steps, I slipped into the nearest row of blue plastic seats. They were icy to the touch. Wind snatched threads of hair from under my hood. I tried to orientate myself against the shifting landscape. There was a blurring between hills, lake and sky. Dusky greys leached into one another ahead, but the boat sailed on darkness. The wooded hillside to the left threw down a blackened reflection that was broken only by occasional fingers of light from the shore.

The elderly man was sitting alone. Where was she?

I scanned the top deck. He was sitting on the boat's right side, mid-way to the front. There was nobody else. A single lamp hung from a pole out ahead of the boat. I thought of a lighthouse. I thought how a lighthouse that moved would be no use to anyone. The rows of plastic seats were covered in shadow. Where had she gone?

In the gloom, I picked out a pair of swans. Swans mate for life. A man had told me once how they liked to remain close to each other, but not to touch. I imagined her fall. It was hard to judge exactly how far above the water we were. Would she have scattered the surface with a splash, or parted it evenly like the cormorants did? I could hear the empty chimes of buoys. No shouts. No screams.

Had he pushed her? She couldn't have climbed over the railing alone. But if she'd resisted he couldn't have got her over. Not so quickly. How could he be sitting there so calmly? The lake was deep beneath the boat.

I thought I should shout for help, or run below deck and ask the crew to stop the boat, but I didn't move. The man's back was inscrutable to me now. I realised at no point had I seen his face.

Lights neared the boat at the water's edge and assembled themselves into a town. It had been too cold for her, of course. She'd surrendered to her age. I'd been distracted by the view as she passed. She was below deck, safely returned to the snug of the bar. He'd remained on the top deck to watch the lake for them both.

The boat cut its engines as it pulled into the marina. Only in their absence did I realise how constant the hum had been. I returned below to the booth I'd left. The air inside was so thick with warmth it was difficult to breathe. My empty cup remained on the table. As the boat came to a stop an automated announcement asked us not to move until it had moored, '... and thank you for joining us on Miss Lakeland today. We wish you a safe journey onwards'. The barman continued to read his paper. Rum dripped into the mug. The young couple both touched their phones. The mother swaddled the baby against the cold. The girls flew about trying to escape their father and the confinement of coats. There was an audible crush of wings. It didn't matter how hard I watched the empty booth. She wasn't there.

Afterword:

Galileo's Ship

Prof. Roman Frigg
London School of Economics

In 1632, shortly after the publication of *Dialogues on the Two Chief World Systems*, eminent Italian polymath Galieo Galilei was summoned to stand examination by the Holy Office of the Inquisition in Rome. In June 1633 he was found 'vehemently suspect of heresy' and forced to abjure.

His heresy consisted in defending the Copernican theory of the universe. The traditional worldview – originating with Aristotle and given a canonical form in Ptolemy's *Almagest* – saw the earth as the immutable centre of the universe. Other heavenly bodies, including the sun, moved around the earth, which explained the pattern of day and night and the observed motion of planets in the sky. Copernicus' *On the Revolutions of the Celestial Spheres* proclaimed the bankruptcy of this view and announced a fundamental reshuffle of the cosmic order: the immutable centre of the universe is the sun, and the earth moved around it. Having attributed to the Ptolemaic view the status of an untouchable dogma, the Church did not look kindly on what it saw as a violation of the injunction against promoting a heliocentric universe. Galileo was forced to repudiate his views:

> 'I have been judged vehemently suspect of heresy, that is, of having held and believed that the Sun is the centre of the universe and immoveable, and that the Earth is not the center [sic] of same, and that it does move. Wishing, however, to remove from the minds of your Eminences and all faithful Christians this vehement

> suspicion reasonably conceived against me, I abjure with a sincere heart and unfeigned faith, I curse and detest the said errors and heresies, and generally all and every error, heresy, and sect contrary to the Holy Catholic Church. And I swear that in the future I will neither say nor assert orally or in writing such things as may bring upon me similar suspicion.'[28]

Legend has it that he acquitted this abjuration with the famous 'eppur si muove' (and yet it moves), declaring that the earth moves around the sun after all. The sentence was sufficiently ambiguous to get away with, but the consequences of the condemnation were severe. Galileo had to spend the rest of his life under house arrest, and *Dialogues* was added to the index of prohibited books (where it stayed until 1835; a formal pardon from the Church had to wait until 1992).

The doxastic reasons for rejecting Copernicus' heliocentric worldview are need not occupy us here. What is of interest, even to the modern reader, are the empirical refutations heliocentrism allegedly faces and how Galileo dealt with them. The argumentative strategy underlying these alleged refutations is to establish that the earth cannot possibly move because if it did there would be all kind of observable effects, which are, however, never observed. Consider a ball dropped straight down from a tower. If the earth was moving at hundreds of kilometres per hour, the ball would not land at the bottom of the tower but far behind it because the tower has moved while the ball was in the air. However, we do observe balls to land at the bottom of the tower when dropped. On a moving earth we also would have to experience constant strong winds, analogous to the draft we feel when driving fast in an open car, and we would see that birds would not be able to fly in all directions with equal ease. Yet we experience no such winds and birds do not fly in privileged directions. On a moving world, grains in a bowl would have to disperse as they would if you if you suddenly started whirling the bowl around your head. But no such dispersion is ever observed: grains remain in their bowl. For these reasons the world must be at rest.

To uncover the flaws in these arguments Galileo invites the reader to imagine the following scenario:

> 'Shut yourself up with some friend in the main cabin below decks on some large ship, and have with you there some flies, butterflies, and other small flying animals. Have a large bowl of water with some fish in it; hang up a bottle that empties drop by drop into a narrow-mouthed vessel beneath it. With the ship standing still, observe carefully how the little animals fly with equal speed to all sides of the cabin. The fish swim indifferently in all directions; the drops fall into the vessel beneath; and, in throwing something to your friend, you need throw it no more strongly in one direction than another, the distances being equal; jumping with your feet together, you pass equal spaces in every direction. When you have observed all these things carefully (though there is no doubt that when the ship is standing still everything must happen in this way), have the ship proceed with any speed you like, so long as the motion is uniform and not fluctuating this way and that. You will discover not the least change in all the effects named, nor could you tell from any of them whether the ship was moving or standing still. In jumping, you will pass on the floor the same spaces as before, nor will you make larger jumps toward the stern than toward the prow even though the ship is moving quite rapidly, despite the fact that during the time that you are in the air the floor under you will be going in a direction opposite to your jump.'[29]

The modern equivalent of this thought experiment is the observation that life goes on as usual even if we are in an aeroplane flying from London to New York at over 900 kilometres per hour: if you drop your glass it lands at your feet, not at those of the passengers five rows behind you. And so on. Galileo's point was striking: the moving earth is like the ship! Just as the motion of the ship is not observable for those on board because all process and motions take place as they would on land, the motion of the earth is not observable for us because motion does not change the way in which things move. In fact, everything inside a moving object – be this a ship or the earth – has the same forward motion and when the ball falls from the tower it in fact keeps moving forward at the same speed as the tower itself, which is why it lands at the foot of the tower and not behind it.

We now know that this argument is only approximately true. Unlike the ship, which is assumed to move on a straight line at constant speed, the earth moves in an elliptical orbit and it rotates around its own axis. This has an effect on the motion of objects on the earth. However, the effects are so small that they are negligible in everyday contexts, and the effects certainly aren't the ones Galileo's critics envisaged. In the tribunal of history Galileo was the clear winner of this debate.

But there is a deeper message in this thought experiment. Setting aside picturesque analogies and cutting to the deep structure of the argument, what Galileo showed is that the laws of nature are the same in a system that is at rest and one that moves at constant speed. This is now known as the *principle of Galilean invariance*. The significance of this insight can hardly be overestimated. It introduced into physics the notion of studying invariances and symmetries, which has become crucial in all branches of modern physics, most notably in Special Relativity. So Galileo's Ship not only proved arguments against the heliocentrism wrong; it also gave physics one of its most powerful theoretical instruments.

Notes

28. Quoted in M. Artigas and W. R. Shea, *Galieo in Rome. The Rise and Fall of a Troublesome Genius.* (New York: Oxford University Press, 2003) p.194.

29. G. Galilei: *Dialogue Concerning the Two Chief World Systems*, trans by Drake Stillman. (New York: The Modern Library, 2001) pp.216-217.

Monkey Business

Ian Watson

In all five directions forests stretch away from the city of Scribe where the 37 robot monkeys type in the Templum daily from dawn till dusk.

Amidst those forests, tended by fairly happy peasants, are pastures and pools and arables to feed the citizens of Scribe. Rivers run through, transporting logs to the paper mills. The whole wide world folds itself in a fifth direction so that no place is really far from Scribe and at the same time, resources are abundant.

Vast, for example, is the Plain of Paper, where here and there checked pages are stored giraffe-high in batches tied with ribbons. Robot giraffes are the cranes of this world. On that otherwise empty Plain of Paper, rain never falls; the only wind there is a gentle, dry breeze.

Betty whistles to herself as she strolls along one of the rides of Forest Seven, approaching ever closer to Scribe. An adventurous lass cannot be satisfied until she has seen the robot monkeys typing. She swings the wicker basket cradling her ploughman's lunch, not that she nor her Dad nor her Mum are plough people, but rather cheesemakers. A lark chants tirra-lirra.

'Why,' she exclaims to herself of a sudden, 'I may arrive on the very Day of the Play!'

She is overheard. A figure steps from betwixt the serried pines – youthful and handsome enough, blue-eyed, his shaggy hair flaxen, clad in goatskin breeches to which goat hair still clings, giving him the aspect of a faun or satyr. His leather jerkin is buttoned by bone. A wallet hangs from the belt of his breeches. His hat, obligatory in public by the Law of Hats, is of floppy felt with a goose feather. He smells fragrantly of lavender, rather than of goat, so maybe his mind is set on seduction rather than on ravishment. Momentarily, nevertheless, freckly Betty fears for her relative virtue, but the youth merely accosts her with words:

'Maiden, the Day of the Play may be a million years away. Or a trillion – at the last syllable of recorded time. We should never *expect* to see a low probability event.'

A muffin cap hides Betty's bundled-up hair since she hates the bother of braiding, although its carrot colour and a few stray strands hint at her hair hue. Over her laced linen vest, a short-sleeved marigold shift and a skirt tucked up to save it from dust and pine needles. Freckled arms, oh yes. Her clogs and his could have come from the same klomper.

Mettlesomely, she replies, 'If the play typed by the monkeys is inevitable some day, due to the Law of Extremely Large Numbers, why must the play arrive almost infinitely far ahead in the series of typings? Why shouldn't it happen in the midst of the series, or even *near the start*?'

'I think,' muses the goatswain, who realistically must be a goatswain rather than a faun or satyr, 'this is a Halting Problem. Though truly I'm unsure. I can't decide.'

These peasants aren't clotpolls. Elementary education flourishes so that everyone may contribute with a will, be his or her condition ne're so vile, and have something beneficial to think about while ploughing or weaving or the hundred other things.

Well, mostly. Always there are some calibans.

With a merry wink the young man adds naughtily, 'Hey nonny no.'

'This,' says Betty, 'is neither the time nor place for any nonny-no.'

'Not on a soft green bed –?'

'Of prickly pine needles! I'm off to Scribe to admire the monkeys.'

'Coincidence! So am I. My name is Orlando. I have saved up a shiny shilling and four silver pennies.' He jiggles his belt pouch.

'Enough for three geese and a pound of raisins –'

'Or three chickens and 66 herrings – ' he responds, since everyone knows the value of money.

'Or 24 tankards of ale, a week in an inn, and two quails –' And Betty grins.

'Maiden, you have just mentioned a week in an inn – how long do you purpose to stay in Scribe? Hast thou kin to stay with? I am no robber, I swear, but hast thou so much coinage as me?' What a knavish lad, presuming to *thou* and *hast* her so soon. Yet he does not smell at all rank, and his teeth gleam. Nor indeed is he a gent of rank, masquerading as a swain to beguile. He seems honest.

As if he can read Betty's thoughts, Orlando looks crestfallen.

'Forgive me. I have told thee a lie. My true name is not Orlando... but *Toby*. "Toby or not Toby?" – I mislike the merriment which my birthname oft prompts.'

'Then you shall be my Orlando.' Betty too can be gallant.

'And together we shall hie us forthwith to Scribe... And we may talk in plainer fashion. We are not mechanical monkeys – consequently semi-quotations from plays by Himself cannot bring the Day of the Play any closer.'

This is true, the whole point about the monkeys being that they possess no language, thus their typing is totally stochastic. (Or is it so, *strictly speaking...*?) No monkey will realise if it gives rise to the whole of *Hamlet*, word perfect. That recognition requires a literate checkernun, who also feeds in fresh sheets of paper, a respected occupation in Scribe; women have patience.

And the monkeys need to be robots, otherwise they would soon tire of banging typewriter keys and start scratching their armpits or their privy parts, and chatter and hoot, or sit backwards on the stools to which they would need to be chained. Real monkeys certainly wouldn't type from dawn to dusk, although this is required.

Let's see: we'll peg the fastest *realistically sustainable* typewriting speed at 50 words per minute, otherwise the keys jam, or fingers fail. Define an average 'word' as five letters long. Thus 'honorificabilitudinitatibus', 'the ability to achieve honours' (*Love's Labours Lost*, Act V, Scene 1) – which might be pompous nonsense, but at least it's *His* nonsense – is equivalent to 5.4 words. Call 'dawn to dusk' 12 hours. 36,000 words. That's rather more than a *Hamlet*, and half as long again as an *As You Like It*. Let's just say for simplicity: a play's length per day. 37 monkeys could type the whole canon during one day. Though that adds probabilities (or improbabilities) enormously. So the law of typing states: *one* perfect Play.

Anyway, a robot's fingers, or its machine, may need cleaning and a spot of oil on its carriage rails. If the checknun-paperchanger wants a pee, or lunch, she signals for a substitute, yet many assorted things can hold up the typing, such as excitement in the Templum at a coherent sentence of text. The peasantry are aware of all this due to elementary education – the overwhelming majority prefer a rural life to the hectic pace of existence in Scribe.

'Ah, but which Play?' teases Betty as she and Toby-Orlando set off together through the sun-dappled pine forest where birds sing like tinkling bells, hey ding a ding, ding.

'Maiden – surely I cannot carry on calling you Maiden so anonymously –'

'My name is Beatrice.' Truth be told, the people back home know her as Betty.

'May I call you *Bee* for short? Since your loose strand of hair is the hue of a honey bee.'

'Where the bee sups, there sup you? Nay! And I am not short at all.' Which is true; she stands – or strides – cheek-high to Toby-Orlando. 'I prefer to be Beatrice.'

'Very well, Beatrice, I was about to ask: Are you a One-er, or an Any-er?'

This is less of a controversial enquiry than it once might have been, two centuries gone by, when life began. Do the robot monkeys need to type *one specific* Play? Or will *any* of the thirty-seven Plays suffice? The Law neglects to specify. Brawling broke out on this issue back then, though not serious bloodshed – everyone was too glad to be, rather than not to be.

Betty still remembers her joy when, on her fifth birthday, her Mum explained how she and Betty and everyone else are living in a simulation within Himself, located somewhere, in a sixth, inaccessible direction. She is *cared for*, not a random circumstance. A high degree of peace and contentment is the law of nature, so that people can dedicate their lives to the typing of the Play, either by living in Scribe itself or by sustaining everything that the city realistically requires. There is *purpose*.

By her seventh birthday Betty is beginning to wonder about the robot monkeys.

'A monkey is like a bent child, hairy all over.'

'But, Mum, children like to *play*, not type a Play all day.'

Worn-out typewriters are recycled to the villages, so Betty has seen one by now. The typewriter factory in Scribe builds identical replacement typewriters by hand, requiring many craftspersons and a sizeable infrastructure. The ribbon factory supplies inked ribbons as well as the uninked ribbons used to tie the checked pages of rejected gibberish, and also fancy ribbons for adornment. Whilst the ink factory... oh what a city is Scribe. The Play's the thing! Without the Play, how can order and civilisation exist?

'The monkeys can groom for fleas and cavort together from dusk till dawn in their compound, apart from the off-time of repose.'

'But Mum, a robot is mechanical, a clockwork... how can it have fleas like a cat or dog or hedgehog?'

'It can still *search* for fleas. The thing is, my little darling, a simulated real monkey would only type spasmodically, if at all. So we must have simulated robot monkeys to type. But the robot must resemble a real monkey closely enough that we can still call it a monkey, such as the law ordains must type, and as Himself provides for the purpose from the beginning.'

We have not yet mentioned the monkey repair shop and spare parts manufactory in Scribe, a place where every wonder of the world converges, often by taking advantage of the fifth direction.

'Mum, is a *mon-key* wound up like a clock by using a *key*? Thus its name?'

'My clever darling! But a real monkey and a robot monkey must share very-similar-tude. Such as a hairy coat glued over the metal body. And a tail that moves. And other similitudes. Very similitudes.'

Such is the talk of peasant families. Sometimes. Awareness of Himself and his plays enriches all life.

An argument of the Any-ers is that there are 37 monkeys and 37 plays. This must be significant!

Yet in what way? Is there in any sense a race between the monkeys, which all have numbers from 1 to 37 painted upon their hairy backs? Is number 25 more likely to achieve a complete Play because number 25 has *already* typed at random a perfect six sentences from that very same Play? So is number 25 less likely to succeed with a *different* Play? Is *As You Like It*, which is fairly short, more likely to occur sooner than *Hamlet*, which is fairly long?

Anyway, the Any-ers fancy that all of the Plays are equally likely and unlikely. Yet all are necessary so that we have a full vocabulary. The One-ers fancy that there is a hidden principle of *primus inter pares* – alias, first amongst equals. *Hamlet* is often mentioned if someone says in an alehouse, 'Suppose one Play alone could survive...' Others advocate *The Tempest* – or *The Winter's Tale*, which may be the reason why robot bears are sometimes spotted. A bear was caught in a pit five years hindwards and proved to be clockwork. The bear was like... a piece of scenery or of furniture. Unlike sheep and pigs and cows and horses and chickens, for instance.

Both factions *fancy*. To believe would be too assertive. Bloody noses might result.

'Of course I'm an Any-er,' Betty answers her Orlando-Toby. 'With 37 chances, surely any one of the Plays is more likely sooner rather than later?'

'Yet what if a Play *not* by Himself is typed perfectly before any play by Himself?'

Briefly, doubt clouds her brow.

'Is that not,' he pursues, 'success of a sort?'

'We'd have no way of knowing if such a Play is authentic. It may arise spontaneously. There may be no original which it copies. *Itself* is the original.'

'And it may seem very like a Play by Himself which, however, he never wrote amongst the 37. So this must be discarded, although tied with a unique red ribbon.'

It's warm in the woodland. Bees buzz. Betty wipes her brow. She planned on a mellow stroll to Scribe, not a dispute en route. However, Orlando's comments are cogent, and almost like a kind of courtship.

'Don't forget the Law of Close Enough,' Orlando teases. 'How close is close enough? We're aware from certain intrusive footnotes – which must never be typed – that the canonical texts which we possess are sometimes a compromise. Oh that this too too solid flesh would melt... or is that flesh *sullied*?'

She sighs. The erotic implications of his words do not elude her.

He continues, 'I cannot believe that such flesh as yours, for instance, might ever be sullied even if its solidity – its resistance – melts.'

Thus, may she yield to him without shame? Even, shamelessly? How his teeth gleam, how his biceps bulge, how his gaze reflects the sky, hue of a robin's egg.

'So therefore,' he declares, 'I am a One-er. Himself chose the perfect expressions of thoughts. We must come as close as possible – to *one* Play, not to 37. Yet what occurs, I ask you, if and when that Play is achieved? You hope for this event to happen today, at random, Beatrice. Yet what then? Will the world simply *halt*? So therefore take thee best advantage of the present time...'

'Huh,' she says, hitching her skirt, the better to step over some nettles intruding upon the path without catching a sting between her legs. 'That's no pretext to *halt* hereabouts – don't you notice the nettles?' So saying, she picks up her pace.

Their first sight of Scribe is spectacular: two broad rivers bridged by masonry, rafts of logs destined for papermaking being poled along on each river amongst the many swans; smoke and steam rising lazily from redbrick edifices which must variously be the pulp mill, the ink factory, the ribbon factory, the typewriter workshop, the great forge and smithy, the central market, the shambles, the monkey menagerie and adjoining repair shop for robot monkeys, for robot giraffes, for coal-fired armour'd robot wagon-rhinos, and for robot logging elephants – although most of the elephants are out in the many forests. And of course there are thousands of garden-girt cottages as well as sprawling tenements and mansions of rank. Many are the docks and wharfs. Yearly, Scribe sprawls further. Soon it may be necessary to introduce public transport, a steam-train to the suburbs, say.

Up over there, topping a central hill, must surely be the marble Templum! A wide flight of twice twenty-four steps rises up to that fourteen-sided Templum, quite like the rows of keys of a typewriter, only twelve times more so – the famous Qwerty Stairs, no less!

Soon enough, Betty and her Orlando are sitting on a bench in a leafy bower beside Coriolanus Bridge, sharing her ploughman's of Cheddar and crunchy barleybread with a nice nutty taste.

'Where shall we go first?' he asks.

'To see the monkeys, of course! Why else did I come?'

'First, I think we ought to investigate the possibility of staying a few nights at a hostelry. Travellers may arrive constantly and occupy rooms. We don't wish to share with four strangers, us all sleeping three to a bed. Those strangers might have bugs and lice on their bodies and in their clothes.'

'The beds themselves may have bugs. That's why I plan to go home tonight to Myrtle-by-the-Water.'

'Fie,' cries Orlando in mock dismay. 'That isn't very adventurous of you. Ahem,' – and he opens his wallet, whereupon a stronger scent of lavender whiffs out – 'you may reply upon me to sweeten the bedding, supposing the maid has neglected to.'

Maid, indeed. Ale-wench, more like, if you're lucky. However, this goatswain is foresightful!

'Besides, a bite of bread works up some thirst in me.'

Within the alehouse, several blokes – who must work in the Ink Factory to judge by the stains on their hands – are arguing about the laws of typing over pots of ale and cakes. Maugre the law, one cannot pause to describe all their hats.

'If only the entire text could be in upper case!'

'But that would not be the Play *itself*, Jonas.'

'Listen to me, chum, using the shift key randomly introduces an entire extra multiplication of improbabilities. It's as if Someone Up There doesn't *want* our monkeys to succeed.'

'At least there ain't no italics key – !'

A red-haired chap with a bulbous nose breaks in: 'Monkeys, plural, cannot succeed – one monkey *only* must come up with one complete entire Play all by itself. Supposing each of the 37 monkeys produces several different perfect pages from the same Play during the same day, these pages cannot be summed together.'

'Excuse me,' interrupts Orlando, 'but won't that satisfy the Law of Near Enough? If this happens during the same dawn-to-dusk?'

Indignantly, from Jonas, 'I very much think *not*, Bumpkin.'

'If thee dub me Bumpkin, I may stick thee with a bodkin!'

'Peace, everyone,' calls out the alekeeper, who probably keeps a wooden rolling pin nearby.

Orlando takes the opportunity. 'Mine Host, is a room free in thine inn? For the usual tuppence? With lavender betwixt mattress and sheet?'

'Free for a price, aye. Sans lavender, tuppence. Lavender for one penny more.'

''Tis fortunate that I brought my own lavender.'

As Orlando reaches for his wallet, Mine Host calls, 'Nay, no need to show lavender!'

'Nor,' murmurs Betty, 'thus to sweeten the sour ale...'

'Will the mattress be of swan's down, or of straw?'

'Straw? What do you take the Queen's Head for? The mattress is likely of soft hay! For tuppence more, our best room boasts a mattress and bolster of chicken feathers.'

'Feathers sound soft,' remarks Betty. 'Hay is too close to home.'

'Alas, our Best is let already out – to a famous actoress who must not blemish his complexion.'

'A *hay* and a ho,' quips Orlando. 'It's settled. And no sacks of chaff for pillows.'

'And no *nonny*,' Betty murmurs, though in a naughty tone, maybe due to the ale.

'May I see the room?'

'Nay, not yet,' she protests. 'The monkeys!'

So it seems they will spend at least a night or two at the Queen's Head. The actual Queen's Head, of brass and clockwork, is inside the Templum, as official head of state. Wound at cockcrow, she recites the law of typing while an amanuensis huddles by candlelight comparing the received parchment in case of any changes in wording. A town-crier stands attendance, just in case of any change. The Queen's Head is basically a mouthpiece; administration as such is in the hands of a privy council.

As for the touring performances of the 37 plays by the Queen's Men, exact copies are of course mandatory, to be scrupulously checked by proof squires, then certified by a chamberlain who embosses a seal upon each approved page. Worn and torn copies are added to the flames on Forks Night when everyone toasts manchet buns of twice-sieved wheaten yeast bread flavoured with rose water, a rare treat.

As everyone knows – unless they are a poltroon, or a caliban.

Betty has seen Plays enacted on the village green of Myrtle-by-the-Water a score of times by now.

'Good rustic, give me your pennies to bite,' says Mine Host. 'As is my custom with every guest *in posse*.'

Is this a sly insult? What peasant would dream of counterfeiting silver pennies by baking tiny hardtack biscuits, strong as iron, then painting them silver, using white lead, say? Just for instance! Surely naughty money is an urban legend.

But Orlando assents gracefully.

'Soft enough,' judges Mine Host, and pockets the coins.

Betty, indeed, has two whole silver shillings in her scrip within her basket, though that isn't the sort of thing to divulge to an Orlando whom she has scarce known for a couple of hours.

The inkmakers already returned to their conversation. A William is saying, 'To produce capital letters and small letters in italics as well as Roman will require four possible settings for the shift key – or, I suppose, two shift keys with different purposes... And each typebar will require four characters mounted upon it, namely upper and lower case italics *as well as* Roman. All for what? In the text, italics are used for personal names as well as for stage directions, but *never* for *emphasis*. So I do not think that the law seeks to handicap our monkeys. Why, the handicap *could* be doubly so.'

'Nay, *quadruply* so.'

'Nay,' from another voice (and hat), 'handicapped twenty-fourfold squared times twenty-threefold squared times twenty-twofold squared and so forth. Methinks. Without even taking into account punctuation.'

'Would that be factorial or exponential?'

'Faugh, that none of us is a mathematician!'

Exeunt Betty with Orlando.

Betty and Orlando climb the Qwerty Steps, hand in hand – 'Lest she slip,' says he gallantly. The Steps are marble so that they may endure for ages, each step engraved at each side with one of the keyboard letters, including punctuation, though no italics or space. Best carrara marble from the old quarry away off in the fifth direction *that way*. In the early years of existence the quarry was a hive of activity – never nearly so much since, although citizens can still obtain slabs for patios if they're rich; hiring a robo-rhino to haul a sledge costs a pretty penny in coal from the mines over that other fifth way.

Talking of hives, beeswax gives the Steps their sheen. The beehives of the Templum are off in yet another fifth direction, surrounded by thousands of acres of white clover. Where the bee sups. Sucks.

Talking of coal, the climate's usually mild with hot sunny spells, as you like it, though now and then there are Lear storms when the rain it raineth every day, and some Winter when icicles hang by the wall. In the countryside the peasants burn wood, the forest wardens keeping an eye upon their takings since paper supplies must never be put in peril. Within Scribe itself, it wouldn't do to burn wood, which is paper *in posse*, consequently coal is best to burn even if a bit costly, and even if much lumber must become pit props to obtain the coal.

'The Qwerty is not so slippery,' protests Betty, scarcely attempting to shake her hand free from Orlando's grip.

No sooner has Betty so declared, than an unsightly

beldame – hook-nosed, squinty, almost a man's moustache upon her upper lip – comes caterwauling atop the Qwerty clutching a wet bundle.

'Curst be the drowner of my cat! Mine Ariel, God's lioness! Aye-aye-aye-aye!'

A rope-hooped farthingale, lacking any overskirtle, juts out around her already wide hips. Her stiffened clothes spread wide. She's like some upside-down ship, full sail inverted, as she collides in her derangement with the leathern bucket at the side of the topmost step, spilling a wash of water.

Such buckets occupy the upper flight of the Qwerty, buckets of water to the far left, of sand on the right, part of the fire brigade protecting the Templum in case of an overturned dawn or dusk candle conflagrating. Throw sand on to any burning paper, not water, needless to say.

Just behind the beldame appear a constable, to be addressed respectfully as Master Justin or Constable Case – the arm of justice, *just in case* of any criminal activity or civil unrest; plainly the beldame fits the latter category. The pimply fellow, clad in pumpkin pants and hose, is brandishing his truncheon.

'Hold, Mistress, hold!'

Another bucket of water tips right over, cascading, and Orlando is quick to tell Betty, 'Into my arms! Let me carry thee, coz.'

Already fire-brigade spillage is reaching her feet.

'Dolt, I'm wearing *clogs*!' quoth Betty.

'Too much of water hast thou.' Whereupon Orlando catches Betty behind the knees then under the back, and he hoists her, her feet kicking somewhat. He's strong, and he mounts the remaining wet steps while the constable belabours the beldame's hooped farthingale with his stout stick – this won't cause her much harm beyond a few bruises but should nonetheless tame her. The beldame loses her bundle. Wits unhinged, her flailing hand disarranges the slashes in the Constable's full pants so that his codpiece juts prominently, policeman's protector 'tis said; quite a trick to protrude it through trunk hose. That raggy bundle of hers unwraps itself from step to step, a shroud emptying out a sodden dead mog.

So here are Betty and Toby at last, where the robot monkeys type, clickety-clackity times 37 in the white marble hall.

Strictly speaking, these monkeys might better be called baboons. Though what's in a name? A baboon, could he speak, would own a name, and thus comprehend words, *which may not be* according to the law of typing.

The monkeys sit on three very long benches at three very long tables, a typewriter in front of each. Their hands never stray from the keyboards to play with themselves nor scratch an itch. The great brazen keys in their numbered backs turn too slowly to perceive.

Close on a double-score of wimpled checkernuns feed blank paper, scrutinise vigilantly, stack typed pages for pageboys to bear away to rearward tables for clerks to ribbon. Some constables circulate. Betty and Orlando are the only visitors in the Templum just now.

Ding, sing the line-ends, *ding ding ding ding*.

'Some may deem me a bumpkin,' says Toby, 'yet I ne'er thought till now: since candles abound, why may not these monkeys type thorough the hours of night equally as day?'

'Because the monkeys need to wind down?'

'Aye, that may be...'

'In sooth, verily!' A constable loiters near them. 'Ahem, a separate night shift of monkeys was *not* provided *ab initio*. Necessitating, by the way, a noctural troupe of checkernuns and clerks, as well as –'

'– a complete ocular industry *ab initio* to supply eyeglasses for checkernuns with increasingly bleary eyes?' guesses Toby.

'Consider the risk of fire!' Whereupon, the constable continues his patrol of the Templum.

Hmm, verily. A sufficiency of giant candles might only increase the acreage of white clover for bees quarterfold above domestic use, but what of an increased risk of fire in the Templum, all be it now of marble...?

Toby gestures. 'Coz, see, the Brazen Head's over there –'

Just then, a checkernun calls out excitedly, 'Text! And more text!' A thrill runs through the whole hall as the Marshall of Pages – majestic in tall crowned swan-feather hat, frilly ruff, velvet doublet, and knitted hose, his whiskers well trimmed – strides to see the typed page emerge, and clerks converge. Other checkernuns should not leave their posts, but do stare.

'I told thee, Orlando! 'Tis the Day of the Play!'

Fortunately the event is happening at the front table. Although a constable extends his truncheon to bar Betty and Orlando from getting too close, their young eyes are keen as eagles', to behold:

> oT vqP ?aaaa gHiMb, bzEYq !, gFistOOO? nnHrr gAAA Amleto: Essere, o non essere, questo e il dilemma: se sia piu nobile nella mente soffrire i colpi di fionda e i dardi dell'oltraggiosa fortuna o prendere le armi contro un mare di affanni e, contrastandoli, porre loro fine? Morire, dormire… nient'altro, e con un sonno dire che poniamo fine al dolore del cuore e ai mille tumulti naturali di cui e erede la carne: e una conclusione da desiderarsi devotamente. Morire, dormire. Dormire, forse sognare.V k !rteW ,hHle zAfs R

The Marshall slaps the monkey's left shoulder in case it continues typing sans paper.

'What text be this?' exclaims a clerk.

'Be this the language of the Turks?' exclaims another. 'Of Illyria? Or India?'

Of course one knows of such places, since Himself knows, even though such places are in no direction.

'Mine eye may be deceiv'd,' declares the Marshall, 'yet methinks this be the language of Verona or of Venice. Sir, or not Sir... quest... dilemma. Nay, but mark: *Amleto*! Amulet, omelette... nay, *Hamlet*, misspelled! Thus I deduce that this partial text represents 'To be, or not to be, that is the question', in Venetian or Veronian!' He raises his gaze roofwards. 'Himself, art thou translated?'

When existence first began, the Templum was of oak beams and joists, the typewriters already in place, the robot monkeys seated waiting for the clong of a matins bell, the *First Folio* for checking, a sufficiency of blank paper. As is known from the *Chronicle*, to all but poltroons and calibans. That same *Chronicle* which first was progressively penned, using print-ink, in the wide margins and on blank versos of *The Law of Typing*, partner to the *First Folio* – until a printing press and bookbindery was up and running, after mining and metallurgy and forestry and paper production, which the first people took to almost instinctively. Truly, that was a Golden Age of inventions and initiatives. The first simpeople, forefathers and foremothers, were far from being simpletons! And they knew they were *sim* – *not* simian; that was for the monkeys – since they could have no false history prior to the start of typing, otherwise the start of the first day of existence, at three of the clock in the afternoon, would not have been a true start. Aye, at three post-meridian – yet what meridian was it *post*? Nothing to be seen in the dark backward, from the posterior of that first day; prior history, a blank! Yet these initial value constants of the clock and calendar were givens.

Yes, three post-meridian was when sentience began. Barely time to get adjusted before the matins bell of the following morning commenced the first day of typing.

True, the start is somewhat veiled in obscurity due to space constraints in *The Law of Typing*. For many years the first people were all too active as ants for a scribe to set down progressively fallible memories. That first generation, children included, were a bubble of the earth.

The first form of the Templum was inadequate even

though it would have made a fine mansion. Dust drifted down. Splinters dropped to lodge between the keys. Mice and spiders soon invaded. More dignity, please, for the *raison d'être*! Busy as beavers, the first people saw to that dignity within a few years. They felt compelled, yet free in most respects to choose how they achieved their goals – as if an invisible Reeve was presiding, supervising the work on Himself's property.

Perhaps the world was scanty to start with, but how quickly and logically it developed towards the best conditions for the Play, as though by chaos plus necessity. Pulp mills, ink factory, robot logging elephants, an entire economy. Some of it built with the sweat of the brow and biceps, some of it bodying forth to requirement like airy shapes solidifying – which latter does not so often happen these days, though happen it does.

'Shall I summon the Town Crier,' a clerk asks the Marshall, 'for him to call out this text from the top of the Qwerty?'

'Fie, how to pronounce these foreign words? *Essery o non essery kwesto...* The words may be correct, or they may be inaccurate. How could we know if 'colpi' should be 'culpi'? We *cannot* adjudicate this text. And is a translation in any way legitimate? At best 'tis a distant version, not *First Folio*.'

'O horror! horror! horror!' exclaims the checkernun. 'Tongue cannot conceive this text nor voice the names! Mayhap a hundred versions of Himself exist in unknown tongues. When I took my vows of vigilance, I verily believed that our tongue is the preferred target of the apes amidst the vast majority of gibberish. No, not of the *apes* as such, nor the target as such, but the clear preference of the Play, whichever it may be.... Tush, I express myself poorly.'

'Sister, apes have no tails.' The Marshall points at the long bench. 'Thereby hang many tails.'

'What price,' muses a clerk, 'a perfect version in a language we know not, nor never will, unless there be some clue such as 'Amleto' whereby we can deconstruct and reverse engineer the language in order to check? Furthermore, I hazard that in futurity languages may arise which have no existence yet. Would a perfect version *typed now* in a not-yet-language be legitimate? Indeed, what is gibberish now may one day be vindicated as valid. In the long run.'

Betty squeezes Toby's hand in excitement. This seems unlikely to become the Day of the Play as such, yet to participate in such bestowals of wisdom is a privilege! Not to mention the thrilling alarm of the checkernun. Life is quieter in Myrtle-by-the-Water. Nothing wrong with quiet, mind thee. Meantime the other 36 monkeys are continuing, *clickety-clackety, ping ping ping ping ping ping...*

Toby ventures upon a hug.

'Surely,' says the Marshall, 'we can tell much about validity from the distribution of spaces. For instance, capitalV-space-k-space-space-space-exclamation-rte-capitalW-space-c-comma-h-capitalH-le-space-space-space-z-capitalA-fs-space-space-space-space-capitalR,' pause for breath, 'is unlikely to be legitimate in any language. Not that our forefathers and foremothers came into being with much ken of any language other than English, merely awareness that other languages are possible.'

Is Toby's hand caressing the clad side of her breast as if inadvertently, merely a part of the hug? Knavish lad! If only a couple of his fingers could reach as far as her still-clad nipple.

'I adjudicate,' the Marshall proclaims after a while of reflection, 'that this *substantial* fragment in Veronian or Venetian be classified *blue ribbon*, and kept in mine office hereafter. Yet it cannot be deemed canonical, not being in Himself's own English.'

'So, coz: much ado about *nothing*,' Toby whispers warmly in Betty's ear. Can he be alluding yet again to her nonny, with which he must be aching to make a lot of ado?

'Fie, you make me flush.'

'Like a bird with a bush?'

'Fie, for shame...'

'Fee-fi-fo-fum, slip on your bum, down topples she.'

'How puckish thou art.'

Perhaps reluctantly, they turn their attention to admiring the oaken table, guarded by two constables, whereon stands the Queen's brazen head, and on which repose, bound in brown leather and chained, the *First Folio*, the *Law of Typing*, the *Book of Probability*, and the *Law of Hats* which covers everything else; a hat's at the head of a human, topmost item we turn to.

'For a sixpence,' a constable confides, 'I shall open one of the books.'

'Nay, nay... thanking you,' says Orlando.

'The same sixpence, I shall slip into the Queen's own mouth, and she may tell your lass's fortune. Or not.'

Betty smirks. 'Mayhap I already know my fortune.' At which, Orlando exhales.

After they have gawped for a while, in chorus: 'Let's see more sights!'

Merely one of which is an elephant lumbering along a paved street well wide enough to accommodate the log which the beast bears in its trunk – though, were the pachyderm much taller, that log would collide with the machicolated timbered upper floors which lean towards one another.

'Make way, make way!' calls the driver perched upon the beast's shoulders, for this is a street of shops displaying trinkets, toys, and gaudy things such as attract crowds. Coal smoke pours from the chimney on the beast's back.

That elephant may be a blessing to distract Betty's eye from the shop frontages and thus conserve Orlando's coins...

By now much time has passed with this and with that. Her basket already contains some souvenir trifles. Roofs have hidden the declining sun.

'Look!' she exclaims. 'Yonder inn sign is of an elephant too!' Some hundred paces ahead.

'Perchance a sign to us, to pause at that inn on our way to our hostelry?'

'Hardly a coincidence, Orlando. Such beasts may sometimes walk this route – not every day nor often, else shops might lose trade. The beast and the inn sign are linked, not two items conjuncting as if by chance.'

Betty remembers her simple petty school lessons on probability better than he. Or him. Her grasp of grammar is fairly good. The dame who teaches in Myrtle-on-the-Water studied at the Faculties of the University of Scribe for a year and has a good memory as well as several books. Dame Polly Pomfret even makes lists of words as Himself spells them, and their meanings, some of which still need to be deduced. The greater part come unbeckoned to the understanding.

Of a sudden, rapidly rising in volume from a shadowy side lane, comes a hue and cry of '*Caliban! Caliban!*' As shoppers rush together in solidarity, there stumbles from that lane a mockery of man, a finny black figure spotted with creamy blotches the size of half crowns. A little dog yaps at the creature's heels, such as those are.

'Take shelter behind me,' Toby tells Betty, bunching his fists.

'True to its name, the terrier terrifies the poor Caliban – see how it gasps.'

'I don't think 'tis made for running.'

The Caliban's face is fishy, as though it belongs in shallow water amongst reeds along a river. Bulgy-eyed, it gapes one way then another, unable to spy escape, especially when in one direction an elephant looms, though less loomingly by now.

And now Constables arrive in their pumpkin pants, leading a chase out from the lane. One brandishes a hoop of rope on a strong long stick. Hue and cry dies down. Soon enough the Caliban is hooped, the rope drawn tight.

A Constable announces, 'To the river with it, where it belongs. Who wishes to come, may follow, to watch it swim away.' The creature is hauled off, trailed after by half a dozen excited townsfolk.

Others of the posse spy the sign of the elephant further

along the street – the receding elephant too, although that's less momentuous than a halfman-freak loose, no longer in the streets.

'Come, coz, before the inn crowds.'

The Elephant inn is much larger than the Queen's Head, boasting a cobbled courtyard, galleried at first floor level.

Two stools soon accomodate Toby and Betty at a shared tressle of drinkers and eaters in a capacious oak-beamed room.

'That pie looks good,' Toby declares of the mountainous meaty wedge on the platter of a red-faced trencherman opposite to them. A well-dressed gent, his garb much trimmed and embroidered. His skinny companion is tucking in to what remains of a whole roast chicken. Napkins are tossed over both men's shoulders.

'I heartily recommend the pie,' munches the gent, taking a swig from a leather mug. 'Kidneys and oysters and coney, well herbed, a fine combination.' Indeed, sucked rabbit bones litter the side of his trencher, near to his feathery hat.

By now a dozen men and women are pushing in, chattering animatedly about the Caliban.

'May I ask, Goodman and Goodmaid, if something sensational happens outside?'

'Why yes, Your Honour, a Caliban is caught.'

''T'is strange,' says Betty, 'I know what they are, yet at the same time I scarcely know not what they are in essence.'

The skinny fellow sucks a chicken leg bare, discards the bone, swirls his greasy fingers perfunctorily in a nearby basin of water, wipes with the napkin.

'Those things,' he asserts, 'are errors. Or glytches.'

Toby is glancing around at other drinkers and diners. 'I see no pots of ale anywhere.'

'Why, Goodman – '

' – Orlando – '

' – this is no common alehouse, but a tavern. No thin potations here. My name is Burgess, by the way. I take it you do not stay here at the Elephant, coming I suppose from the countryside?'

'That is right, Mister Burgess.'

'*Errors*?' asks Betty, having no idea what might be a glytch.

'We stay at the Queen's Head, should you know it.'

'The Queen's Head? That, I should describe as an error. Have you tested their beds yet? Those are by no means tester beds! Is the Queen's Head even licensed for rooms?'

'We are humble folk, Mister Burgess. We need not four posts and a canopy.'

'Faugh. Are you on your hony moon, Goodman Orlando and Goodmaid – ?'

' – Beatrice. Nay, we met today.'

'You deserve better than an alehouse. Innocence should not be so abused.'

Betty nudges Toby. 'Hearest thou?'

Toby glances again for wench service, in vain. The nearest wench is forever looking a different way. However, Mister Burgess booms out, 'Claudia!' Within only a few winks, the wench attends them.

'Ah, Cloddie, two sherry sacks for my friends.'

'And two pieces of that same pie,' adds Toby.

'Nay, a salad for me, good Claudia,' says Betty. 'I feel this is my salad day.'

'Food from the ground is lowly,' observes Mister Burgess.

'Boiled carrets, scallions, radish, sparagus, coucumbers?' the wench recites, and Betty nods, adding, 'but with not too many spices.'

'You are considerate, coz, regarding my coins.'

'Oh *I* shall pay for this, dear Orlando.'

'You shall pay *dear*?' enquires Mr Burgess with a wink.

'*Errors*,' she repeats, eyeing the chicken chewer. 'How so?'

'You may call me Master Morgan. I profess at the University. Calibans are the embodiment of *coding* errors in our world.'

'*Codding* cues for lechery,' observes Mr Burgess merrily. 'Whatever *coding* may mean, I must scratch my head. At times Master Morgan waxes so, his brow lost in the clouds.'

Claudia deposits two leather mugs in a hurry.

After her first gulp, Betty says, 'Why, 'tis so sweet and rich. How it warms my blood.'

'Yet in truth it is watered here – as I like it, I confess. Imagine the prince of wines, of which this is a bastard. The vineyards of Sherris are quite far off in a fifth direction.'

'Why, Your Honour, there seem so many fifth directions.'

'Thus our world ever reaches out and enricheth itself. Consider the oysters within my pie. An oyster requires a sea. A sea implies a ship. A sea may merely circle around, yet must be salty. Truly, this world of ever more fifth directions is our oyster.'

'Yet may,' enquires Master Morgan, 'such growth continue without cease?'

A great moist wedge of pie arrives, along with Betty's salad, and they tuck in while Master Morgan continues, still heeding his remains of roast fowl, 'Is our state finite? Suppose our world to be a machine, like a typing monkey, but made of noughts and onces...'

The parson's nose is especially succulent; some save it till near the last of a chicken.

'Any machine with a finite memory must have a finite number of states, thus any pre-ordained menu implemented upon this machine must either eventually halt or repeat itself. The duration of our pattern cannot exceed the number of inner states. Yet the number of inner states may be as the life of an elephant to the life of a mayfly.'

'Or,' continues Master Morgan, 'tearing off a roast wing, is our rhythmic sequence non-preordained? How to tell the difference between one seeming near-infinitude and another?'

'Thou art a philosopher,' says Orlando around a mouthful of pie, and signals to Claudia for another Sherris sack.

'If only philosophy could find this out... And what is *time* for Himself, compared with for ourselves? Maybe our own lives are as of mayflies even if we seem to achieve three score years and ten, thus to allow millions more opportunities for the Play to occur. On the other hand, the requirements for running the task expand multifoldly more than the task itself, occupying ever more processing space.'

Mister Burgess cuts in abruptly. 'Hang up philosophy! Goodman Orlando, what dost thou pay at the Queen's Head?'

'Tuppence. Mine Host already bit my coin.'

Mister Burgess sighs. 'Methinks a bed at the Queen's Head is worth but a halfpenny. Here the regular, licensed cost is a tuppence.'

'Oh.' Does Toby flush with embarrassment, or due to the Sherris sack?

'A tuppence to tup in comfort... And the pillow will not be a sack of chaff. Good rustics, I wish you both very well. Let today be your hony moon, here... If 'tis good 'tis done, 'tis best 'tis done sooner, say I. As to your tuppence which the aleman took, mine brother-in-law Ralph happens to be Scribe's official ale-taster.' A position of rank! 'Ralph can pass by the Queen's Head on the morrow. Be assured that Mine Host will restore your coins to you. A conner, as we call the ale-taster, keeps watch for any illegitimate conduct by houses that brew their own ale.'

'I know not how to thank you,' begins Toby.

'Nay, you may drink my health. Though not too many times – I am more accustomed to Sherris sack than thee. Is this not, Master Morgan, a consummation devoutly to be wished?'

'Methinks the lady may protest...'

But Rosy-cheeked Betty is all smiles. 'Oh brave world, that has such people in't! Indeed, thou art no Pandarus.'

Outside, the light waneth much more. 'Tis evening. Night draweth nigh.

The moon does not shine bright in such a night. Mister Burgess's influence runs to a lanthorn to help the anticipative couple navigate the many dark narrow corridors of the Elephant till they reach their bedchamber as instructed. Toby hangs the bright lanthorn by the bed.

'Let us satisfy our eyes.'

Although Betty has nothing to worry about, she teases: 'Some charms are best seen by candlelight alone, sans the reflectors of a lanthorn.'

'Nay, I would not singe thy pretty hairs.'

Toby's member is stiff with desire, and Betty is fulfilled many times, fully filled and fulfilled climactically, yet ne'er does Toby gush. Probably that's the fault of a little too much of the Sherris sack... and maybe of the novelty, the unfamiliarity – Bett's cunny may not feel at all like Toby's own spittled hand.

Toby does not come to climax, no nonny no. Famously ale – thus, how much *more so*, sack? – increaseth the desire but taketh away the through-flow, namely the ejaculate. Toby does not provide through to completion. Per*forate*, yes, with vigour – yet finally per*form*, ah nonny no. He certainly tries. To Betty's delight. For a half hour or more he's cock-in-her-hoop.

'Oh Orlando – Orlandoooo! Do – doooo!'

All's Well That Ends Well.

Mayhap.

Afterword:

The Infinite Monkey Typing Pool

Dr. Rob Appleby
University of Manchester

THE IDEA OF A ROOM full of monkeys ceaselessly bashing typewriter keys until an entire Shakespeare play is produced is derived from a very real mathematical concept: the Law of Truly Large Numbers. This law puts into words something that we all find every day – surprising and rare events do actually happen – or as the law states, given 'a large enough sample [enough monkeys and enough time], any outrageous thing is likely to happen'. This idea has been understood by statisticians for decades,[30] but was first put through rigorous scrutiny by the mathematicians Persi Diaconis and Frederick Mosteller, who considered a series of seemingly outrageous coincidences (like the case of a woman who won the New Jersey State lottery twice in four months):

> 'The point is that truly rare events, say events that occur only once in a million [...] are bound to be plentiful in a population of 250 million people. If a coincidence occurs to one person in a million each day, then we expect 250 occurrences a day and close to 100,000 such occurrences a year.'[31]

In some sense this goes against the meaning of the word 'surprising' – the mathematics says we should actually expect hundreds of thousands of surprising events to happen every day around the world. I would argue that it's only our humanity – the powers of imagination that have evolved in us only to deal with small sample sizes – that prevents us from understanding that this is the way of the world!

The Lottery offers us the perfect example. The probability of winning the top prize in the UK's national lottery is about 1 in 14 million. Small. But every other week some lucky person wins it. This is because many people buy a ticket, each of which has an independent chance of winning the top prize, and so the chance of the 'unlikely' event happening in at least one of these trials is much more favourable.

As Diaconis and Mosteller pointed out, psychology also plays a huge part. Not only is it impossible for our imagination to process a narrative with all 14 million lottery players in it (narratologists suggest the optimum number of characters in a story is more like 11 or 12!), if we divide this story into many, we are also more inclined to pick up on a positive story than a negative one. So the story of the one-in-a-million winner rises, improbably, in our imagination, above all the other 14 million stories of minor disappointment. Thus the story of the national lottery is the story of the winner. This goes a long way to help us understand gambling in general, particularly problem gambling, where the lack of understanding of chance ('Well, someone has to win'), and the tendency to forget losses (or loss narratives) can lead to terrible spirals of addiction. The human brain is not well equipped to deal with large numbers, at a great cost to many.

In fact, the human brain is not well equipped to deal with a lot of things that, mathematically, are quite straightforward. Like complexity, for instance. If we fail to get our heads around most things in medicine, it's perhaps because, in that field of science, so many outcomes are the result of multiple factors combining and interacting in extremely complex, if not chaotic, ways. Our brains cannot process this, especially when the outcomes are negative. When something tragic happens, we tend to plump for a simple narrative: *if only I had done this; if only the doctor had done that...* Simplify it, introduce a single 'turning-point' moment (or better still a villain!), and suddenly we can understand it again.

So, back to Ian's story of robot monkeys bashing away in the Templum. In mathematics and science, the monkeys are a metaphor to help us understand the law of large numbers. Or at least try to. If we imagine a single monkey on a single typewriter, he or she will press keys at random and write gibberish. But we could imagine an 'a' following an 's', followed by a 'u' and so on, giving the meaningful word 'sausage'. This is not meaningful to the monkey, of course, and is the same as the last seven successive and random letters. We can figure out the chance of this happening based on typing speed and number of keys but it adds little to the discussion, apart from throwing around some large numbers! Eventually a play will be typed, given enough time. This is the outrageous event in our idea of very large numbers.

The image of monkeys typing away happily provides an insight into what infinity means, or at least gives us a practical definition. It's hard for us to understand how improbable this task is: if every atom in the universe was a monkey on a typewriter we can estimate, based on typing speed, how long it would take the universe to write *Hamlet*. The length of Hamlet is the problem – the number of letter combinations increases exponentially as the play length increases (that is to say much faster than proportionally) – and the monkeys would need a time much larger than the ultimate age of the universe to type every combination of letters. In other words, in a typical universe they would start at the big bang and keep typing until the universe ends, without success. (Maybe we should just appreciate what the Bard did!) Of course, these monkeys may get it right first time, in the first hour, but then again I may also win the lottery this weekend!

This fascinating idea appears a lot in popular culture. In one episode of *The Simpsons*, Mr Burns shows Homer his room of 1000 monkeys working at 1000 typewriters, scolding one of them, somewhere harshly, for writing 'It was the best of times it was the blurst of times'. Indeed, inspired by this Simpsons episode, an American computer programmer, Jesse Anderson, has created a suite of virtual monkeys, using Amazon's cloud computing, to spew out sequences of nine letters (to give you a sense of the scale of this project there are 5.5 trillion different combinations of any nine characters from the English alphabet).[32] At time of writing, Anderson is still waiting for the complete works of Shakespeare, even by accepting any incomplete fragments of prose! But it's fun to try.

When they eventually get round to *Othello*, Anderson's virtual monkeys (and Ian's robotic ones) will need to type the lines: 'And O you mortal engines, whose rude throats / The immortal Jove's dead clamors counterfeit / Farewell!'

The phrase 'mortal engines' has been picked up by a number of science fiction writers. Shakespeare probably meant it to mean 'the canons of war', but our monkeys are free to interpret it an infinite number of ways – for instance, as that strange hybrid, the 'living machine' that is to say, *us*. For the life within that machine is perhaps the ultimate 'complex system'. Imponderable. Something mere numbers, no matter how large, can never measure.

Notes

30. The origin of the thought experiment is a 1913 essay by Émile Borel, 'Mécanique Statistique et Irréversibilité' ('Statistical Mechanics and Irreversibility'; *Journal of Physics*, 5e Série 3, 189–196) in which he argued that if a million monkeys typed ten hours a day, it was extremely unlikely that their output would exactly equal all the books of the richest libraries of the world; and yet, in comparison, it was even more unlikely that the laws of statistical mechanics would ever be violated, even briefly. In *The Nature of the Physical World* (1928), the physicist Arthur Eddington borrowed this comparison saying that the likelihood of a particular event in an experiment was less likely than 'an army of monkeys were strumming on typewriters' writing 'all the books in the British Museum'. How Shakespeare

entered the thought experiment, exactly, is less clear.

31. P. Diaconis. & F. Mosteller, 'Methods for Studying Coincidences', *Journal of the American Statistical Association,* Vol. 84, No. 408 (1989).

32. http://www.bbc.co.uk/news/technology-15060310

Equivalence

Sandra Alland

THERE'S A SILKS AERIAL trick where I climb two hanging lengths of fabric right to the ceiling, then wrap the thick ribbons around my waist and hips like I'm a human yo-yo. I hold my breath and the audience holds its breath, and then I let go into freefall and unravel all the way down, ten metres – whoosh – stopping with a jolt twelve centimetres from the stage. My hair, when it's longer, sweeps the floor for extra drama. No harness. No safety net.

The crowd always goes wild. From their perspective, the appearance of risk is the same as risk. The norms in the audience forget about my unusual body and their uncertainty about my gender. They fall in love with my falling. With them as witnesses, I fall in love too. A person in freefall does not feel their own weight.

The room has only a bed. No bedside table, desk, wardrobe, laundry basket. There's no space for anything but the bed, which is large and almost fills the room. I lie in it or sit on it, but I'm almost never not in it. It's been like this for days or years, I don't know. I woke up and I was here, alone.

The walls are white. The orange is orange. When the shaking stops, I peel. Eat. Drop the rind to see what will happen. It falls to the floor. Sometimes I do this, painstaking and slow, in the dark. I would say 'at night', but there's no window so it could be any time at all. Staring into the dancing shadows, I sometimes think I catch a glimpse of orange peel after it leaves my cramped hands. Orange and floating.

Before this room, I used to ease my pain and insomnia with good drugs and 3am cable. There's no TV here, so I recreate shows in my head. I saw one about Einstein that had a jazz soundtrack. The editors loved overlaying quotes in tacky fonts onto pictures of his shock of white hair and moustache. It was just boring enough to be mesmerising. But when the narrator started talking about Einstein's happiest moment I bolted awake again. It had to do with lifts. Or as Einstein called them, elevators.

...had finally overcome some of the problems with Newton's mechanical model of the universe...re-framed gravity in an exciting and unified way.

Whenever I mentioned the circus, people assumed it was before some terrible accident, or before scoliosis and fibromyalgia destroyed my body. But it was during it all. I've always been a crip. I laugh to myself, sing, 'I am disabled but not invalid!' It's from the opening number of our freak show. I used to love entering the high top on crutches, then leaping onto the silks while the crutches clattered to the floor.

The social model of disability was a wondrous invention, one of the few big British ideas that didn't devastate continents. That's because we disabled people came up with it, not the government. We argued that we're not disabled by some medical notion of being abnormal or less than ('Your spine is twisted! It's a tragedy!'). Instead we're disabled by a lack of lifts, ramps and opportunities, or by the price of electric scooters and drugs that work. But governments never want to provide such things; they prefer we disappear.

I could leave the bed if there were somewhere to go. Or a way to get there.

The floor is painted white, at least what I can see of it. I throw my legs over the side of the bed, poke the floor with my big toes. It's cold and has no give. A cramp rages up my left side. In the corner where the door looms, the concrete is slightly scuffed. The door opens outward and isn't locked exactly; it just doesn't budge when I push on it. It has something that could be one of those eye scanner and thumbprint things on the wall next to it, but I've always refused to let anyone take my data without a fight. I'm pretty certain it wouldn't work if I wanted to try anyway.

The pain is all of the options of the questionnaires: throbbing or sharp; constant or creeping; male or female. Can you feel this? This? Does it hurt in one place or many? But back when I had room and energy to walk, I often walked fine. I was graceful and gliding – no accidental foot-crushing on the dance floor, and glasses of wine were safe with me. Except when everything glitched. Something caught, a flutter, a flurry, and I was face-down.

I began to fall often. Nothing was more wrong with me, I just lost confidence in my relationship with gravity when I wasn't in the air. I suspected the ground. The pavement, the grass, the damp earth seemed to be seeking me out. My hands usually didn't have time to brace or buffer my descent. Especially when I was using a stick. Bam. Flat on my nose with grit between my teeth before I even knew I was falling. Face a bloody mess, knees battered, ego bruised. A quick look around and a hoisting up of the bag of bones, more sore now.

Newton's theory of gravity never held water for Einstein, because one of its implications was instantaneous 'action at a distance'. A picture of an angry-looking Einstein floats in my head, with 'You're wrong!' in Comic Sans. *Einstein pointed out that far-apart objects affecting each other in an instant was impossible, because it meant the objects were sending information faster than the speed of light.*

I stare at the light fixture in the ceiling, alas it's just out of reach. Sometimes I jump to try anyway, and sometimes it does feel a long while before I land back on the bed. The door opens when I do this, and a hand appears. It's a bit creepy and a bit funny at the same time. The hand is likely attached to a body, but from my angle I can be certain of only the hand. There's no way of knowing the rest. The fingers spread as if to say *Stop*, and the hand usually leaves if I sit down.

Jumping isn't very satisfying anyway, it's nothing like flying. And there isn't room to land solidly, so I have to bounce on the mattress. The space around the bed is only twelve centimetres on three sides, and the fourth side is right up against the wall. If I want floor, I have to land on one foot so I can fit between the bed and any of the three walls, and I must always land parallel to the bed and twist my twisted upper body. I managed two feet once, one in front of the other, but I dislocated my ankle. I'm hypermobile and doctors used to say I shouldn't do acrobatics. But a good shove and things always end up back where they belong.

That's what we did in the freak show. I became friends with a Mexican giant, Jesus el Gigante, who was particularly helpful at repositioning shoulders and ankles. He was also a wizard at poetry. He used to recite these haiku-like stanzas, timing the last line with a painful push and a smile as big as my head. Then he'd offer a pull of single malt whisky from his flask. It almost made me want to dislocate something. Here, I do at least get some pills; they appear at regular intervals with the bucket and an orange. But what I would do for one of Jesus's poem-push-pulls.

Though nothing changes, I keep my eyes open for clues. I don't recall a trial. I don't remember an accident or a long journey. What's that? Next to the pile of orange rinds I spot my stick. I climb across the bed, reach down for it. How did it get there? I pick it up, drop it to see what will happen. It falls. Will someone pick it up if I don't, or will it be there forever? If it is in a different place when I next awaken does that mean:

A./ Someone moved it
B./ I moved it in my sleep
C./ I moved it and forgot
D./ It moved on its own
E./ The floor moved
F./ Something else entirely

A while ago I took my clothes off in the dark and dropped them on the floor. They fell too. In the morning, or the next time it was light, they were hanging on the edge of the bed. Last lights out I tried again, and this morning they were piled in a corner. I know they've (been) moved, but I can't be sure if they're clean. I sniff my shirt, my jeans. Pull them on. Now, like when it's dark, there's no way of knowing what spine, what genitals I have. Anything not measurable is not real.

The light goes out. I start jumping on the bed and thinking about Mars. I remember an early-morning TV programme featuring young people talking about the future. 'We should put all the poor people on a rocket ship and shoot it into the Sun', one posh kid said. Replace poor with trans or disabled, and you get the gist of the scintillating variation of the rest of it. But I think over the years it's been mainly rich people who got on rocket ships. They'll soon go to Mars, not the Sun, and sadly not because we're finally getting rid of them. It will be because Earth is fucked. Maybe they've made it to Mars already without letting the rest of us know.

As I jump higher in the darkness, I imagine where a window would be, if this room had one. I imagine being able to see the latest unfoldings of global warming, like maybe actual summer. I wonder how long this can go on before I destroy the mattress, and whether I will get a new one. Who reads all those questionnaires, keeps track of the boxes ticked. Suddenly the hand reaches into the room with a torch. It holds the torch up high, points the light straight into my eyes. I fall. My head smacks against the wall just below where the light is making a nice circle, so there's a red circle too. Pretty.

The hand seems relieved, or maybe distracted. It's going now, the door shuts. I spring up again. I'm jumping particularly well today. My legs are sorer than sore, but I'm getting into the bounce. I feel something close to joy.

I parrot the documentary's BBC accent, 'Einstein eventually rewrote the theory of gravity to include the curvature of spacetime, finally bringing gravity into the Theory of Relativity.

The title 'Special Theory' was replaced with the more authoritative 'General Theory'.

I jump higher, yell in my own voice, 'I get where Einstein was coming from – I've been called special too!'

The door swings open again, and the torch-hand is back. I shout, 'Why couldn't he just let the mystery be!' The torch-hand points the beam exactly as before, but this time when it points the torch at my face the beam of light curves around me and down, missing me entirely. It curves like a giant invisible hand has swooped it out of the way. Or as if something is sucking me down, as I jump, into a circular dent in the mattress that the light has to bend itself around to keep going. I stop jumping and watch the light, try to understand its message. I fall again. I black out.

In rewriting our understanding of gravity Einstein's first step was the Equivalence Principle. I allow Einstein his moment of joy in my head, picture the photo with an animated *Eureka!* jiggling his bushy moustache as he thinks about gravity and elevators. A bad German accent emerges from the photo's superimposed moving mouth:

'You are in an elevator. And you drop your pen or your sandwich or your dildo or whatever. And it hits the floor of the elevator. And subconsciously you feel relieved. You are pissed off that your sandwich has floor grime on it or that the vase with your mother's ashes has shattered, but you are comforted in the knowledge that things are as they should be. You drop something, it falls to the ground.'

Then Einstein ruined everything. He pointed out that you could equally be in a closed room in a rocket ship, and that the rocket ship could be traveling in deep space far away from any large masses that produce gravitational pull. You could be out there in your windowless box, floating, but it would seem like there's gravity when you drop your toothbrush. This would happen if the rocket ship was accelerating towards the toothbrush at 9.81 m/s2, the rate of freefall on Earth. We can experience the mirage of gravity when gravity is nowhere to be found. You'd only notice if the rocket ship changed speed or direction, or passed a giant star.

Under certain circumstances, you just can't tell. It could be that your dropped dildo is being drawn to the Earth's gravitational pull like I've been drawn to ill-fated couplings with a kaleidoscope of genders. It could also be that instead the floor is moving towards your dildo in outer space, speeding up at a precise rate. Your dildo ends up on the floor regardless, and it feels just like home. Like the real thing. And there's no way of knowing whether it isn't.

When I come to, the torch-hand is gone. I can almost swear Jesus el Gigante has been in the room, the smell of his aftershave is that strong in the air. I feel almost certain I heard him whisper:

we look up at the stars
wonder if they also
look up at us

And then: 'Lucha! Fight!' as he did something horrifying but effective to my left shoulder.

It's a long while before the light comes back on, but I don't sleep. When my eyes focus, the first thing I see is my stick, propped against the wall near the door. I reach for it, imagine I smell El Gigante on its handle. Imagine him pretending to use it as a toothpick, laughing.

I laugh too, because now I have a plan. I begin to rip and weave my sheets into two long strips. The material doesn't flow as well, but it does the trick. I jump and hook the pseudo-silks around the ceiling lamp, then tilt the bed on its side against one wall to give me some floor. I quickly perfect all the ooooh ahhhh manoeuvres – the swing, the spool drop, the those-biceps-are-so-big-is-that-a-man-or-a-woman-honey. I wave at a picture of Einstein in my head as I hang upside-down from one ankle; he winks when I blow camp kisses. Is it just the eternal stink of humanity on the other side of the wall, or is it maybe the vast nothingness of outer space? Both are equally possible and equally disappointing. Einstein's happiest moment, my bleakest.

Now for the finale, all wrapped up and ready to unfurl. Like all lifts, Einstein's has given me the ability to get somewhere, even if I can't know where. Back in 1907, he couldn't have known that his greatest comfort would be anyone's daily terror. I only curse him gently as I look down from the highest point in the room. Twelve centimetres is twelve centimetres, regardless of why.

The whoosh fills my ears, and I smile. I remain weightless, whether here or performing for enraptured norms and my darling fellow freaks. Falling is falling, after all. Yet the difference between one and the other is the distance of galaxies.

Afterword:

Einstein in a Lift

Dr. Ana Jofre
SUNY Polytechnic Institute

EINSTEIN'S 'HAPPIEST MOMENT' IS to his General Theory of Relativity what the falling apple is to Newton's Theory of Universal Gravitation. Sitting in his office in Bern, Switzerland, in 1907, Einstein realised that a body in freefall does not feel its own weight; it was a realisation that led to the principle of equivalence, which states that being within a gravitational field is indistinguishable from being within an accelerating frame of reference. So how does this relate to freefall? Well, if you are free-falling, accelerating under the force of gravity, the effect of acceleration counteracts the effect of gravity, which is why astronauts orbiting (accelerating centripetally around) the Earth float about in their shuttle.

As the character in 'Equivalence' explains, the equivalence principle states that if you are enclosed in a windowless box, it will be impossible for you to distinguish between the following situations:

(i) the dildo you dropped accelerates towards the floor because the box you are in is on Earth and the dildo is subject to the laws of gravity like everything else on Earth,

Or (ii) the floor and the entire box is actually in the middle of space, accelerating upward toward the dildo you released.

For the narrator in 'Equivalence', meting out their time within a box, these two situations are indistinguishable. The only thing that is certain is that this box is not in orbit, nor is it hurtling towards a destructive collision with the Earth. If that were the case, the dildo would float and there would be no jumping on the bed. For the narrator, Equivalence – that is to say the realisation that there is no physical way to deduce the physics of the world outside their box – is a rather bleak revelation, perhaps the bleakest. So what, exactly, made it such a happy thought for Einstein?

By 1907, Einstein had already published his theory of relativity, which was successful but incomplete. The 1905 version of the theory is now known to us as the 'Special Theory of Relativity', because it functions only under special circumstances. The moment Einstein realised the equivalence principle was the first moment he envisioned a more complete theory of relativity that also integrates gravity, a theory which came to be known as the 'General Theory of Relativity'.

The special circumstances under which the former theory functions are inertial frames of reference. A frame of reference is the point of view from which a measurement is made. Imagine you are in a car traveling along a very long and straight road. In the frame of reference of the car, you will measure yourself sitting still while the trees and signposts move past you. In the frame of reference of someone standing on the side of the road, you will measure yourself standing still along with the trees and the signposts while cars race past you. As long as no one is speeding up or slowing down or changing direction, then there is no acceleration in these frames of reference – they are what we call 'inertial reference frames'. Einstein's Special Theory of Relativity stated that the laws of physics are equivalent in all inertial reference frames, and that light travels at the same speed in all inertial reference frames. Inertial reference frames are symmetric with respect to one another, and can measure equivalent truths: the speed of the car relative to the ground, or the speed of the ground relative to the car. It doesn't matter which way you look at it, the laws of physics deduced from any measurement in any inertial frame of reference will be the same. Einstein liked symmetry.

Now, if you suddenly push the gas pedal in your car, you and everything in your car will feel a pull towards the back of the car, as you speed up (as you accelerate). Someone existing only within the frame of reference of the accelerating car may surmise that there is a force pushing everything to the back of the car. Yet, the person standing on the side of the road knows there is no such force. The beautiful symmetry of their situation is broken. The Special Theory of Relativity no longer applies: the two frames of reference no longer appear to be subject to the same forces. The accelerating frame of reference experiences a fictitious force.

While non-inertial frames of reference posed a challenge to Einstein's Special Theory of Relativity, it was gravity that caused the biggest headache. Among the most significant problems Einstein found with Newtonian gravity is that Newton's gravitational force is mediated by a single equation that does not account for time between a cause and an effect. Imagine two massive bodies positioned at a great distance from one another. There will be a force of gravity that compels each of them toward one another. Now, if one of the massive bodies suddenly changed its position or its mass (by some kind of catastrophe – a supernova explosion, for example), then according to Newton's equation the other body, positioned at a great distance, will immediately feel a change in the gravitational force it experiences. The immediacy of the effect was problematic for Einstein. In Einstein's view, a change in one body cannot cause an instantaneous effect in another (very far away) body, because information travels at a finite speed. Einstein's Special Theory of Relativity revealed that the universe has a fundamental speed limit: the speed of light. Nothing, not even information, can travel faster than the speed of light. Newton's gravitational theory did not account for the limited speed at which information can travel, so Einstein knew his extended theory of relativity should include an explanation for gravity.

The equivalence principle unifies gravity and acceleration and relativity into a single simple thought experiment. A happy moment indeed.

Returning to our thought experiment in a box. Imagine now that you are inside the box accelerating downwards on

Earth under the force of gravity, in freefall towards your eventual doom. In freefall, you will not feel the force of gravity, nor will you feel the effects of acceleration, so Einstein's Special Theory of Relativity applies here, which means you will necessarily perceive light as moving in a straight line at a constant speed. Now imagine that there is a tiny hole at the top of your box where a horizontally propagating light beam enters. From your point of view, the light beam will travel in a straight line, and will shine a spot on the wall exactly opposite to the hole it entered, where it triggers a detector. (See Figure 1a.)

However, from the point of view of someone outside your enclosed box, as they watch you and your box accelerate downwards under the gravity, the light beam will appear to curve downward along with your fall, triggering the same detector on the opposite wall. (See Figure 1b.)

Figure 1.

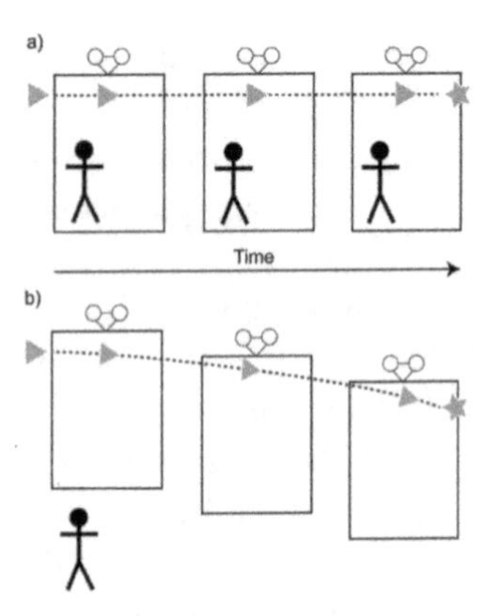

It took a leap of imagination to explain what exactly can universally cause light's trajectory to curve. Einstein's conclusion was that it was the fabric of space and time itself that curved (*spacetime* is actually the proper nomenclature to describe the interweaving of space and time into a single fabric). According to Einstein, spacetime was curved and variable, rather than fixed and rectilinear. At a small scale, within your box for example, you might not notice that the space you are in is actually curved, the way many humans never realised the surface of their Earth is actually curved.

The equivalence principle linked non-inertial frames to gravity. Non-inertial frames of reference led to fictitious forces (such as the backward force you felt when you pressed the car's gas pedal). If the effects of acceleration and the effects of gravity are indistinguishable, then perhaps gravity is not a real force. Perhaps it is a means by which spacetime is curved.

Newton identified objects with mass as the source of gravitational forces. Einstein, instead, identified objects with mass as the sources of curvature in spacetime. Figure 2 depicts the famous rubber sheet analogy in which an object placed at the centre of a stretched rubber sheet creates an indentation in the fabric of spacetime. The illustration in Figure 2 is a useful way to understand how Einstein's General Theory of Relativity explains gravity. An object placed near the indentation will either orbit around the dent (if given exactly the right initial push, horizontally) or it will fall toward the indentation. Light traveling through this region will follow the curves of spacetime. In this view, gravity isn't actually a force, but rather a deformation in spacetime. So objects affected by gravity travel along a straight path, but within curved space.

Figure 2.

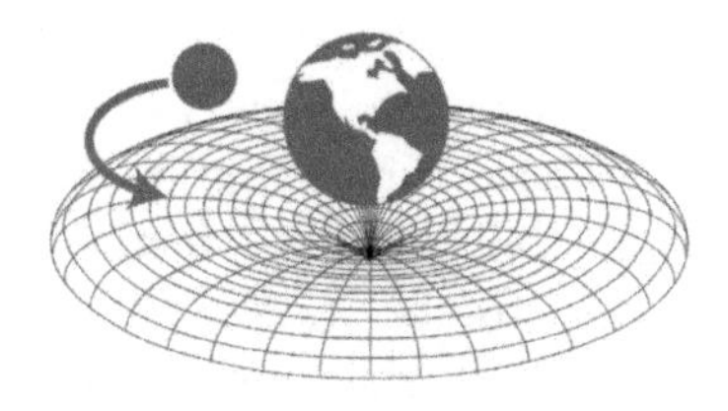

The problem of 'instantaneous action at a distance' is solved when gravity is viewed as a deformation in spacetime, rather than as a direct force between two objects. If a massive body experiences some disruption (such as a sudden change in mass) then this disruption will ripple through the spacetime fabric it occupies, requiring some time to travel across long distances. All ripples through spacetime (or gravitational waves) obey Einstein's universal speed limit.

As elegant as this theory was, a scientific theory needs to be tested against reality. When Einstein published his theory in 1915, he showed that it described the orbits of the planets in our solar system as well as Newton's theory. In fact, the theory of relativity did a better job at describing the details of Mercury's orbit, which was never fully explained by Newton's theory. Further verification of Einstein's theory came in 1919 with Arthur Eddington's experiment in which he observed the position of a star seen close to the Sun in the sky during a solar eclipse. The General Theory of Relativity predicts that a beam of light from the star will bend around the Sun as it passes its vicinity. The solar eclipse presented an opportunity to measure starlight near the Sun. Sure enough, Eddington found that the measured position of the star (assuming the light traveled in a straight path) was shifted relative to its known actual position. It wasn't, however, until early last year, over a century after its original publication, that we finally were able to feel direct evidence of Einstein's General Relativity. On February 11, 2016, LIGO (Laser Interferometer Gravitational-wave Observatory, operated by Caltech and MIT) announced that it had detected gravitational waves, disturbances in the fabric of spacetime caused by two massive black holes colliding into one another. It was a tour de force experiment that originated in the 1970s; its beauty is in its exquisite sensitivity and in the direct tangibility of the evidence it provides for a theory that came from the far reaches of an imagination. We received a direct signal from gravitational waves sending us the message about the catastrophic event between the two distant black holes from which they came.

All this started with a very simple thought experiment. If you were in an enclosed box and you released your dildo…

The Child in the Lock

Robin Ince

THEY WERE GOOD SHOES.

Neil was very happy with them.

They made a statement. They said Neil was his own man.

The shop assistant had told him that when she persuaded him to buy them.

He had never spent so long looking at shoes. A whole day spent wearing out his old shoes by looking for new shoes.

He needed good shoes like he'd never needed good shoes, shoes that were even better than the shoes you'd wear at a funeral.

That good.

He wanted to impress Tom. He wanted Tom to know that he was the sort of person you should be friends with… *had* to be friends with.

Telemarketing does not encourage gregariousness. Fixed booths, fixed conversations, all that talking to strangers made it hard to talk to strangers when you didn't have to. It's hard to sound meaningful when you're paid to not believe in anything you say.

Water cooler moments were difficult because the cooler had been moved to the corner of the kitchenette and you couldn't lean on it without twisting yourself awkwardly.

And it wasn't plugged in anymore, so the water always tasted a bit stagnant.

Tom was the first comfortable conversation he had had in the office. He made it so easy. He wasn't like any other telemarketer Neil had met before. Apparently, Tom had been an actor but decided to take a break from it as he'd always been keen to get into telemarketing, and he knew it was a young man's game.

They'd been talking about Game of Thrones when Tom said, 'You should come round to dinner.'

No one had ever asked Neil to dinner before, not like that, not while talking about Game of Thrones.

There was a hint of flirtation.

Not sexual flirtation, obviously.

No, charm, that was it. Tom was charming. He knew how to look at people.

He was married with children, or maybe a child. You weren't allowed photographs on your desk so it was hard to be sure.

It had been a very long time since Neil had really wanted to be friends with someone.

This would be a good friendship.

They had arranged the meal for the Saturday after next, this gave Neil some time to work out how to make an impression.

He went on wine websites. He'd seen a film once that mocked Merlot, so Merlot was out.

He liked the sound of Pinot Noir. It would sound good when he gave them the bottle.

'I brought a bottle of Pinot Noir.'

Pinot Noir.

PEE-no Nwahr.

That would sound better than Cabernet Sauvignon.

He didn't think it mattered if it was cork or screw top, not nowadays, but he ordered cork anyway.

It makes wine more of an event.

And then he looked at his shoes.

They looked cheap.

Too cheap.

They didn't have a 'hey, who cares about shoes' recklessness about them, they just looked like shoes for people who didn't know about shoes.

The rest of him was just uneven enough that stylishness didn't hang well off him. His walk somehow crumpled everything as he moved, but his feet were good.

His mum said he had good feet because he was a breech birth.

They'd notice good shoes on him.

He arrived at the train station too early. He knew it would be best to be about three minutes late, neither rude nor needy, but he had got to Tom's stop a couple of hours early. You know what trains can be like: well, they weren't like that tonight.

The shoes looked good in the lamplight, not too showy, but noticeable, definitely something you'd want to comment on. These were the sort of shoes that would make you want to become firmer friends with someone.

Neil looked at the Google Map and decided he would stop off at a pub, probably just for a soft drink.

He didn't have a book, he thought it would look clumsy to be carrying a book, and what if it was the wrong sort of book? He'd looked at a few books on the internet, but none of them seemed certain enough. There was always a two star review somewhere or a comment about it being sexist or weird or boring.

It had been a dry winter, so he decided he'd walk along the signposted towpath. He had a handkerchief and one of those special cloths that can shine a shoe with one wipe in case anything messed them and he rarely looked up, so he'd notice any oncoming dog nastiness.

It was much quieter here than where Neil lived. Pretty too, you could breath in without being worried about what you might swallow. He'd like it here. He was playing soft rock in his head, but being careful not to swing the carrier bag in case that spoilt the wine somehow.

Lost in The Alan Parsons Project, he didn't notice the noise at first. The crunch of gravel masked it.

It wasn't an owl.

He listened. Stopped The Alan Parsons Project in his head.

Splashing.

Duck splashing?

Goose splashing?

Hopefully not geese; violent buggers, geese.

He looked down into the canal lock.

It was dark, no artificial light here, but there was definitely something pretty big, bigger than a goose.

There was a boy… or a girl… definitely a child.

What was it doing down there, this wasn't a good place to go on a Saturday.

He moved a bit further and nearly tripped on a discarded scooter, almost scuffed the shoes.

The water was low.

The splashes bouncing off the bricks.

What should he do?

Maybe it was just a lark, they do larks in the countryside and this was nearly the countryside, it wasn't Birmingham anymore.

Neil looked again.

If he went with his gut instinct, he'd definitely think the child was drowning.

But gut instincts can be wrong, he'd seen something about that on BBC2.

He didn't think the child could see him as it splashed, so this gave him further time to think.

If the child is drowning, what would be the best way to stop it from drowning?

There was a metal ladder built into the brickwork, would climbing down to the water help?

Probably... possibly... but what a dirty ladder, dirty and slippery.

A sodden spider's web glistened on the ladder rail.

That wouldn't do his shoes any good at all.

What would Tom and his wife, possibly Jo or maybe it was Hannah, think of his shoes if he had bits of pondweed and mud scabs on it? He'd trail the mud in the house, or have to take his shoes off and he was not fond of the socks he was wearing, he hadn't thought about his socks enough beforehand. Socks had not been part of the plan.

What about opening the sluices so that the water level would rise? Then the child could just swim over and he could reach across to pull him out... But the lock gates looked dirty too.

And what if he ruined his shoes just for someone's lark? The child wasn't splashing so much now, so maybe it was having fun after all. Perhaps it was diving for coins?

He looked down one more time. Why wasn't anyone else on the towpath, it was very unfair to leave this all up to him?

What if the child was drowning and he saved him, and then it turned out that this was the child that would grow up to be a genocidal dictator or something.

He looked at his shoes and he looked at the arms still flapping about.

He'd go to the pub and have a think about it.

It had been a very trying few minutes and a gin and tonic seemed like the best thing to do.

Neil didn't know if he had done the right thing, but at the moment the barman came over he just said, 'Gin and tonic,' without even thinking.

The gin might help him think about what to do about the child. He knew that by doing nothing at least it meant that he can't have done anything wrong, because he hadn't done anything at all.

He didn't hurry the drink, he knew he could only have

the one. It was a nice pub, like the ones you see in Sunday night detective thrillers.

Neil decided it was best not to go back to the lock, whatever happened had happened.

If there was a child struggling in the next lock, then he would get involved, because that was how the balance of the Universe worked. There were no other struggling children the rest of the way, that made him think that the other child probably hadn't been struggling after all, it was a statistical improbability taking into account all the other locks.

He was about to start playing some Creedence Clearwater Revival in his head, but thought it would be best to put them on pause and start thinking about his doorstep icebreakers.

He'd seen a film that everyone liked recently, it was about a man who was a twin and a bank robbery, that would be safe ground and demonstrate that he wasn't just sat at home all the time.

It might even suggest to Tom and Jo, or Hannah, that they should all go out to the cinema together one evening. He'd also found out that the road they lived on was named after someone who had been big in the East India Company in the late 18th century. He'd say something about that as it would show he was interesting like those people on TV panel shows. He started playing Midnight Special loudly in his head.

He was still 11 minutes early, so he sat on a steel bench in memory of Polly Hodson 1927-2005.

The shoes barely looked walked in at all. They were strong leather, perhaps too strong, as they had cut a little into his left ankle. He could feel the light dampness of blood or a hastily formed and burst blister on the wool of his sock. He took out his 'one wipe and they shine' cloth and gave the shoes a light clean.

Lovely.

He knocked on the door at precisely 7.33.

'Pinot Noir!'

Why did he do that? He was far too quick in holding up the bottle of wine, he looked too proud. He hadn't even said, 'Hello'.

Tom invited him in. He didn't look as charming as usual. You don't have to be as charming in your own house.

Neil was right about their being at least one child, there was Lego everywhere and a blue bike leaning on some banisters.

Jo, or Hannah, was on the phone.

Tom told him to make himself at home, but didn't offer him a drink.

Neil worried that he'd smelt the gin.

Did he think Neil was an alcoholic, especially the way he thrust the PEE-noh Nwhar at him.

No one had noticed his shoes yet. He could smell peppers. He didn't like peppers. They gave him stomach cramps, but it was Sunday tomorrow, so that didn't matter.

'Sorry if it seems a bit frantic, we're just trying to find out where Charlie is.'

Neil didn't know what to say, so chose to go with, 'KIDS!' and a smile and a shake of the head.

Jo, or Hannah, came into the room.

'He's not at Teri's or Maria's.'

She didn't look at Neil, 'I'm sorry about this...' She looked as if she was going to continue to explain, then stopped.

'Tom, we should go out and look for him.'

This was embarrassing.

They weren't really taking any notice of Neil at all, trust him to come on a day when they had something else on their mind, it was his usual bad luck.

'Why don't we do this all another time, you've obviously got other things on your plate tonight.'

'He might come back through that door in 3 minutes,

why don't you just sit here and wait. I am sure we'll find him before long.'

'Honestly, Tom, it's not a bother.'

'Thanks for being so understanding.'

And that was that.

The bottle of Pinot Noir was still on the table in its carrier bag. Did he take it with him or leave it there? No one was looking and it seemed unfair to come all this way for nothing. They were hurrying on their coats. It was getting cold now. He'd noticed the canal was beginning to freeze on the way there.

What a disappointing night, neither of them had said the slightest thing about his shoes.

Sitting on the delayed train, he wished he'd bought a screw top bottle.

Next time he saw a child splashing in a lock, he'd probably do something.

As he sat back, waiting for movement and listening to the arguments of couples, he caught a tart and disgusting smell. He'd trodden in something. He took out his handkerchief.

Afterword:

The Spider in the Urinal & The Drowning Child

Prof. Glen Newey
University of Leiden

Robin Ince's short story 'The Child in the Lock' responds to not one thought experiment but two. The first is, strictly speaking, more of an anecdote than a thought experiment, related by the philosopher Thomas Nagel in his book *The View from Nowhere.*[33] In it, Nagel describes a series of encounters with a large spider in a Princeton University urinal, from whose gutter the spider seems unable to escape. Through the summer, the spider survives, even thrives, despite being urinated on 'more than a hundred times a day'. Finally, Nagel takes pity and helps it climb out of the trough with a paper towel. The next day Nagel finds the spider, exactly where he had left it, dead.

Nagel uses this anecdote to demonstrate how, sometimes, our moral decisions over whether or not to intervene in certain situations can be based on questionable assumptions. Nagel had presumed that lifting the spider out of its apparently miserable situation would help it, when actually the golden shower it toiled under was its lifeblood.

The second thought experiment Ince's story responds to is often referred to as 'The Drowning Child', and was first proposed in Peter Singer's article 'Famine, Affluence and Morality', published over a decade before Nagel's anecdote.[34] In his article, Singer puts forward a quite different argument about intervention. Rather than being reluctant to act in order to help others, Singer argues that there is an abiding moral imperative to intervene to help those less fortunate. His argument is that people in wealthy countries have much more extensive obligations to relieve poverty elsewhere in the world than people in the West typically assume. While this is in part a political or public policy question – what percentage of GDP should, say, a G20 nation set aside for overseas aid? – a parallel question can also be asked of individuals. In pressing this latter question, Singer drew an analogy that gave the article its enduring polemical force.

Suppose that while out on a walk you come across a child drowning in a pond, he asks. Nobody else is nearby: you are the only person in a position to rescue her. As Singer emphasises, rescuing the child would come at some cost to you: it would mean getting wet, dirtying and perhaps ruining some of your clothes, and so on. Still, most people would say not only that rescuing her is a good act, but also that failure to do so, assuming you can rescue her, would be morally wrong. Singer invites readers to infer that, by analogy, similar moral wrongness attaches to those who could alleviate poverty overseas (which often has lethal effects) but refrain from doing so.

In 'The Child in the Lock', the narrator Neil faces a very similar decision: he has bought new shoes and when, on the way to have dinner at the home of a work colleague, he sees a boy drowning in a canal lock, he has to decide whether to rescue him, thereby dirtying his shoes. Ince succeeds in making the moral ramifications clear, while also rendering Neil's failure to rescue the child psychologically plausible. While Ince makes clear that Neil's inaction arises from worrying that he will dirty his shoes, the story shows his mental states as being more complex and interesting than just that. By extension it gives a credible insight into the phenomenology of self-deception.

Often fiction handles this phenomenon more successfully than do philosophers, for whom it is puzzling how self-deception can even exist. The puzzle arises from the fact that self-deception seems to call for contradictory attitudes. Somebody who deceives herself doesn't simply get something wrong: she believes something despite the fact that she knows, or at least has good grounds for believing, that it is false. The puzzle lies in the fact that while the attitudes in play here seem contradictory, most people think that self-deception exists. In the story, Neil is not represented as thinking consciously: 'I don't want to wreck my shoes; therefore I won't rescue the boy.' Instead he disbelieves the evidence of his senses. He persuades himself that the boy isn't really drowning, but diving for coins. The psychological realism is reinforced by further thoughts that encourage inaction: that it is unfair that he should shoulder all the burden of rescuing the boy, that he may grow up to be a genocidal dictator, and so on. Neil even has a version of the thought prompted by Nagel's spider story: that well-meaning intervention may do more harm than good. Part of this realism lies in the fact that these thoughts are not entirely consistent. If the boy is not drowning but diving for coins, then Neil's thought about preventing, by inaction, his growing up to be a dictator no longer applies. Neil then medicates his anxiety by going to a pub and ordering a gin.

Immediately after deciding to do nothing and resume the journey to his colleague's house, Neil thinks that he will rescue the next child he comes across who is drowning ('that was how the balance of the Universe worked'), but of course

> '[t]here were no other struggling children the rest of the way, that made him think that the other child probably hadn't been struggling after all, it was a statistical improbability taking into account all the other locks.'

What is going through Neil's mind here? He is casting about for ways to reinforce his earlier hunch that the child is not drowning. Neil's reasoning seems to go along the following lines: the number of locks without struggling children in them has proven to be large; at most one of them has had a struggling child in it; therefore it is statistically improbable that even this one child was struggling. Again this tortuous passage of reasoning seems psychologically very plausible. Neil elects to think he has not seen what he has seen on the grounds that subsequent events have made it statistically anomalous. In his case, the solution to the puzzle facing him – how to avoid a belief that he has good grounds for thinking is true, namely that he has ignored a drowning child – lies in ignoring those grounds in favour of others that suggest the belief is false.

As with most self-deception, this is clearly a motivated belief, in that Neil wants to think that he has not seen a struggling child (which would then lay moral claims on him of the sort Singer discusses, and which Neil has ignored). But the reasoning itself is not entirely without merit: it resembles David Hume's argument against miracles,[35] which goes roughly that you could have adequate ground for believing reports of miracles only if you thought it more miraculous that such reports were false. When faced with a putative 'black swan' event, Hume suggests, the very fact that it is a singularity gives grounds for doubting its veracity. Neil's reasoning proceeds on a similar basis: he hits on a class of event (passing a canal lock) with regard to which the struggling child is a singularity. The problem facing his, and Hume's, reasoning is that there is no accurate way of gauging the base rate. One sign that the reasoning has gone awry is the fact that the alternative event that Neil wants to believe has happened – that the boy was splashing but not drowning – is equally anomalous.

Ince's story raises, in short order, questions in ethics, philosophical psychology and the justification of belief. It suggests that in tackling them, reason may be not so much inactive, as ineffective.

Notes

33. Nagel, Thomas, *The View from Nowhere* (Oxford: Oxford University Press, 1986), 209.

34. Singer, Peter, 'Famine, Affluence, and Morality', *Philosophy and Public Affairs* 1:3 (1972), 229-42.

35. Hume, David, *Enquiry Concerning Human Understanding*, Section X.

Keep It Dark

Adam Roberts

> 'I need so little: a bottle of ink, a speck of sun on the floor – and you.'
>
> *Vladimir Nabokov, writing to Vera Nabokov.*

On the drive over they had a conversation about darkness. Kay said: 'the proportions in the actual universe are 4.9% ordinary matter, 26.8% dark matter and 68.3% dark energy. And actually the majority of ordinary matter in the cosmos is also dark, since the stars that we see and the nebulas and the other bright stuff is only about a tenth of regular mass of everything.'

And Broome replied: 'Fascinating, fascinating, so visible matter is *one tenth* of *five percent* of the whole picture. One tenth of one twentieth.'

And Kay said: 'Yes'.

'How wonderful that you have all these precise numbers at your fingertips!'

'It's my job.'

Broome thought about it for a while. He never sounded more like a preacher than when he replied: 'So, when God said let there be light, he was actually saying, let there be a half of one percent light, and as for the rest, keep it dark.'

'Hmm,' said Kay, navigating the car through a tricky turn, off the asphalt and onto an overgrown gravel track.

'Doesn't that make you wonder,' Broome said, staring through the passenger's window with his sightless eyes, 'if we've got it the wrong way around? The universe is not a blaze of light in the darkness, a big bang, a frame fitted for us to clamber upon.

The universe is dark and it had always been dark and the stuff we're made out of is a vanishingly rare aberration.'

'Spoken,' Kay said, 'like a blind man.'

*

They had gone as far as the D-Max could take them, deep into the wilderness. Kay parked up and helped Broome out.

'Not much further,' she told him. 'I can see the scope.'

They were on a low, broad ridge. To their left the land dropped towards the lake. A little way ahead, to the right, a lesser slope revealed the discarded giant's bra-cup of the telescope in amongst all the heavy greenery.

The air was hot as steam. Sweat kibbling the skin of the back like insect bites. The scalp moist and itchy, skin rashing-up under arms and between breasts, a sheer horrible devilry of heat.

Kay took Broome's hand and lead him. 'Off we trot,' she said. The gates were fuzzed with brambles, jammed ajar by weeds. It was a tight squeeze getting through. Down into the underworld, golden bough in hand. Gold shines only when light shines upon it.

'Off we trot,' Kay had said, but they walked slowly rather, picking their way through the undergrowth.

There were as many mosquitos as there were stars in the sky, and they swirled in great galaxies of dust-mote darkness over the skein of the lake, and sank down as a cloud into the fern-littered valley, as were monstrously curious about Kay's neck and face and arms. Tufts of grass Mohican-sprouted. An ancient tree put its many elbows out at awkward angles. A chubby spider was fiddling with its web. It looked like it was beckoning to them.

A minute brought them to the dish. The squat tower was chained-up with vines like an escapologist's trunk. An oval pond of pure scum nestled in the concavity. Underneath, swallows had built their nests out of dirt and moss and discarded wiring. A congregation of yawning tribbles.

'Wow,' said Kay. 'That's some disrepair.'

'You didn't expect,' panted Broome, '*pristine*.'

'No but,' said Kay. 'Well, but. Wow.'

Broome put his head back and his unseeing eyes did their freaky wiggle-wiggle thing. 'I can hear birdsong,' he said.

'Welcome swallows. Lots of them. Me, mostly I can hear mozzies.' A distant high-pitched drone. A pigmy bagpipe. She was far from convinced that the cream was working. 'Let's get inside.'

The old accommodation consisted of three rectangular pre-fabs and a generator surrounded by a metal fence. The buildings were a grey that was the colour of rain. The windows were all marked with smallpox patterns of dust. Kay couldn't see which one Lorenzini was staying in, but they were all deserted.

Here was a black watertank. It plinked like a tapped bell as insects landed on the drumskin surface of the water it contained.

Here were fat gumtrees with intricately fractured bark and saddlebags of greenery.

Here was a tent, and Lorenzini opening the flap like a surgeon peeling back skin, and out he stepped, looking like the walking dead, but waving to them cheerily.

'You came, you came, you came,' he said, and wrapped his boa arms constrictingly around Kay's body and kissed her cheek. He was weeping. That wasn't a good sign.

'Why are you in a tent, Lor?' Kay asked. 'Wouldn't one of the huts have served?'

'Don't laugh, he replied, gravely, 'but I've lined the inside with foil.'

'Wow,' she said.

'Sure,' said Lorenzini. 'Sure. Better than a hat.'

'Dear lord my friend!' said Broome, looking a yard to the left of Lorenzini and laughing. 'You've gone off your trolley! All alone out here, and fruity as a cake.'

'Rude,' said Lorenzini. He sounded actually hurt.

'You're underweight, Lor,' said Kay. 'Mate, we need to get you back to civilisation.'

'Sure, sure, I know, I know. And you, bat-man!' Now he lunged forward, and threw his arms around Broome, and hugged him tight, and Broome kept laughing, and then Lorenzini was saying: 'You're right, of course, you're right. You're my only friends in the whole goddamn country – sorry, Rev.'

'You have the arms of a gibbon,' said Broome, extricating himself. 'And something of the same smell. Can we go inside? This *crème* isn't keeping these gobbling monsters off my skin, I think.'

And now Lorenzini was wiping the tears out of his eyes with grimy fingers. 'You gotta see my data first,' he insisted, and took a dainty hold of Kay's elbow, as a man might hold an uncooked egg with only the tips of his fingers.

Back up the overgrown path he led them, past the three empty accommodation prefabs, back towards the dish. 'You got data out of *that*?' Kay asked.

'Paradox busting data,' said Lorenzini, with pride, but then he was weeping again. The lines in his face were deep and clear, like the cracks in the ice on the surface of Saturn's moon Europa. He really did not look well.

'Mate, I know radio telescopes,' said Kay, 'and this is not a functioning radio telescope.'

Lorenzini was at the little door in the stubby little tower, unlocking it. 'It doesn't function as a conventional radio telescope,' he was mumbling, or at least Kay thought that's what he was saying. 'But it still takes in data. Come inside.'

'Oh wow,' said Kay, the opposite of enthused, as she looked through the door. 'Nothing is supposed to smell like that.'

'Yeah,' agreed Lorenzini. 'There may have been some lunch-related decomposition. Hey, Rev! Woo-hoo. Hoo!'

'I'm blind,' said Broome tetchily, 'and not deaf.' Kay guided him towards the door. Lorenzini was gabbling on.

'You know Olbers' Paradox, right?'

'Wow,' said Kay, stepping away from Lorenzini. 'Is that a dead *cat*?'

Lorenzini felt the shift in temperature passing through the

door. It was still hot, and the smell was genuinely unpleasant, but it was better than the overwhelming heat of outside. Lorenzini's voice was close by his right ear.

'The paradox is how *dark* the night-sky is. It's a paradox in three senses. For one, Olbers wasn't the guy who first formulated it, it was a geezer called Digges, a century earlier. Two, the paradox is itself. And three, *I've* solved it, so it's not a paradox any more, which makes it paradoxical to *call* it a paradox.'

'I don't think that last adheres to the proper definition of the term *paradox*,' said Kay, further away.

'Is there a chair?' Broome asked. His own breathing sounded horribly close and rough to him. He needed a rest.

Lorenzini guided him along the wall to a hard little seat, and Broome lowered himself down.

'In an infinite universe,' Lorenzini was saying, 'with an infinite number of stars, and well, here we are.'

'Wow,' said Kay again, and not in a delighted way.

'Wherever we look in the sky there ought to be a star. We ought to see stars everywhere, wherever we direct our eyeline there should be a star. The night sky should be brighter than day. It should be always bright. We should never be bothered with night.'

'Kindergarten stuff,' said Kay, dismissively. 'What's this data you were talking about?'

'Obviously I'm not an expert,' said Broome, wiping his face with a handkerchief. Golly but breathing was hard. 'Presumably there's all sorts of dust and planets and whatnot, getting in the way.'

'Interstellar dust, yes, yes, yes, but in an infinite universe the light would fall on the dust for an infinite length of time, and the dust would heat up and start to shine. You gotta remember, infinite numbers of stars. Infinite time. Everything should be alight. Hot and bright and shiny. Bright as midday sun on chrome.'

'So you've got me,' said Broome. 'What's the solution to the paradox?'

'The universe has not been around an infinite amount of time,' said Kay, from across the room. 'It started with a Big Bang. We can't see further, in light-years, than the universe is old, in years. Only finitely many stars can be observed within the light-cone observable from Earth, and that finite number is too low to light the whole sky up.'

'But,' countered Lorenzini, 'if that's true, then all our lines-of-sight should lead us back to the Big Bang. And that was presumably very, very bright.'

'My solution to Olbers' Paradox is who cares. That's my solution.'

'Don't be like that, Kay!' Lorenzini sounded genuinely hurt. 'You're my two best friends in the world! I couldn't tell anyone *else* my solution to the paradox. I couldn't publish it in a scientific paper.'

Such snorting. Much mockery.

'Why don't we,' Kay drawled, 'drive back to town and have this conversation in a nice bar, with a nice cold beer, instead of in this –'. Then she screamed.

There was a flurry of sounds, a clatter and a series of thuds, but most of all was Kay's voice wailing a soprano vibrato. Broome tried to get up, slipped, and went down painfully onto one knee. He put out his arms and touched the wall behind him, and as he did so the door to the little space slammed. Everything trembled.

'Kay!' he shouted into the darkness that was always his darkness, and always there. 'Kay? Kay?'

'Jesus fuck Jesus Christ,' Kay screamed. 'He's thrown something!'

'Kay?'

'Ow he threw something in my eyes. Oh my God, Broom-o, I can't see. I think it was bleach. I'm pretty fucking sure it was bleach in my eyes. Oh my. Wow.'

He stumbled his way towards her, and caught his hip painfully against something hard and unyielding and shaped like a coral. 'Oh my poor Kay! Joe? Joe what have you done?'

'He's gone, Rev. He's gone out and locked the door.'

Broome found her and hugged her, and she sobbed for a while. 'Man, fuck, it *hurts*,' she said.

'I've never known him this bad,' Broome said, and immediately felt foolish for saying so. I mean, *obviously*.

Eventually Kay stopped crying. She groped her way about the space. 'Thirsty,' she said.

He felt his way about the space too, and touched some pretty dubious-feeling stuff, but there didn't seem to be anything to drink. 'You'll know your way about a telescope in the dark better than I will,' he said.

'I never worked this scope,' she said, through gritted teeth. 'What's he playing at?'

'We need to get you to a hospital. If it was bleach he threw – well, they may be able to salvage something. The lenses and the retina will be fine, though the cornea may have some scarring…'

'Rev,' Kay said. 'I know you mean well, but shut up.'

They sat apart. He wanted to console her, but holding her made them both hotter and more bothered. Who knows how long. A long time. The heat and the thirst made Broome torpid.

The sound of the lock opening was a rasp from the end of days, and the next thing the air was full of Lorenzini shouting something incoherent and Kay screaming and then the sound shrank and the door shut again, and Broome was alone.

He prayed. He tried to pray, and did his best. The words formed in his head, and he reassured himself that God was everywhere, even this hellish little space. He reminded himself that Lorenzini was his friend, and not evil – unhinged, clearly, and evidently distressed and certainly dangerous; but still a soul precious to God, and still his friend. Soft words turneth away – turneth away – soft words.

The despair was always there, inside his soul, like the black rock inside an avocado. Soft and green and rotting, his soul. He cried. But then he stopped, because he was a dry old man, thirsty, his innards desert dust, and he didn't have the moisture to spare. Even though every inch of his skin was slippery with sweat.

He may have dozed.

After an indeterminable amount of time the door opened again and Kay was returned.

'Kay! Kay!'

'Oh Christ, Rev,' she said, and there was a gurgle in her voice, like something had come loose in her chest. 'He's gone, he's gone. He's barely even Joe any more. He's lost.'

'He's lost his mind,' rasped Broome.

'He's still got his mind. He has a whole new theory about Olbers' Paradox, and its clever. He's lost his humanity. He said he was in love with me.'

'Kay – no.'

'Right? He's old enough to be my – like in decades, decades of friendship I never got that vibe off him. Not once. But he says he's been sitting out here, in his fucking foil-lined tent, thinking about me, and obsessing. He used that word.'

'Kay,' Broome breathed.

'That's why he called me. He wanted to share his insight. He wanted to throw fucking – to throw into my eyes some fucking – oh dear God it hurts raw, like acid. Maybe it was acid.'

'And me?' Broome asked.

'It's all tangled up with God. Rev, Rev, you got a phone, right? You got some special blind-man's model of a phone? We gotta ring for help.'

'I don't own one, I'm afraid,' said Broome. 'But they're not designed with blind people in mind, mobile phones, that's true.'

'Joe took mine off me. The crazy bastard. Man, we have to get out. There must be something we can use in here to bust the door open – and then –'

'I don't think two blind people stumbling around the Northern Territories on their own are going to fare particularly well,' said Broome. 'Assuming we could get free of Joe.'

Kay fumbled around the space for a while, and found a metal spar, or length of pole, or piping or something hard enough to use as a weapon. With this she smote the metal walls, and generated a huge gong-like din and a shuddering reverberation in the air, until Broome begged her to stop with his hands over his ears. She was just letting off steam, he knew. She was just angry and frustrated and staving off the despair.

'Next time he pops his fucking head in here,' she warned.

They were silent for a while.

'What was his solution to the paradox?' Broome asked.

For a while Kay didn't reply. Then she said: 'He thinks dark matter swallowed the light.'

'Oh,' said Broome. 'Is that – likely?'

'Not really.' She hummed a tune for a while: Broome didn't recognise it. Then she said, 'Though, you know, nobody really knows what dark matter or dark energy *is*. That's why we call it dark. So it's not impossible.'

'But haven't there been experiments with both? The High-Z Supernova teams and so-on.'

'We know very little, really. Like you said in the car: what can we see? One tenth of one twentieth of what's out there. It's next to nothing. Joe thinks that dark matter started out as something else, widely but not uniformly distributed, and that when a photon strikes a particle of this whatever it is, it becomes fixed, loses its luminosity – its speed, really – transforming that into the dark matter equivalent of mass. He says he has the equations to show it, but when he started on those I zoned out. My eyes are fucking sore – and the skin around the eyes is...'

'I don't understand why he thinks this is such big news. Why he doesn't want to release it to the world.'

'The 95% of the universe that is missing is all the absorbed light from the big bang. If that substance weren't there, sponging up the surplus light, the light sky would be 1900% brighter. He thinks it's the devil. I think that's what he thinks – he wasn't very clear.'

'The actual devil?' asked Broome.

'You regretting his confirmation now, are you, Rev?'

'God still loves him.' Broome said, as if reminding himself. 'And He hasn't abandoned us.'

'I'm going to... to... to bash his fucking skull with his bar,' Kay muttered. 'Then I'm going to retrieve my mobile, and then you and I are going to figure out, somehow, how two blind people can make an iPhone work well enough to call the emergency services.'

Broome dozed again. The next thing he heard was a scraping sound, like a needle in the playout groove of a record swinging round and round, earth about sun, death about life, a czzzch azzzch sort of sound. It took him a moment to understand that he was listening to Kay choking.

He stumbled towards her, and hands – Lorenzini's, of course – shoved him hard away. He cracked back against the wall, and there was a sharp pain in his side, and it didn't dampen. If anything it got sharper, harder, more agonising. When he felt the afflicted area he discovered his shirt had torn, and the skin underneath was bleeding. Blood slick with sweat. There was a protrusion of some kind on the wall behind him.

'Joe!' he yelled. 'What are you doing, Joe? Joe leave her alone!'

Nothing.

'Kay? Kay are you alright?'

Nothing.

Broome felt his way around the space, hoping to put a hand on Kay and feel that she was still breathing. But he couldn't locate her.

'On his right hand, all the light from the big bang,' said Lorenzini, over to the right. 'On his left, the devil, the devourer of light.'

Broome rushed at him, and felt a painful blow on his right shoulder. He slid and hit the ground, moaning. Lorenzini had struck him – with the bar, was it?

'Joe, Joe, what have you done? Joe, call an ambulance. We need to get Kay to a hospital. Joe, it's not too late.'

'You never saw her, Rev,' said Lorenzini, dolorously. 'You never saw how beautiful she was, with your eyes.'

'Joe! Joe come on, now. It's not too late!'

'Divide the universe into a series of concentric shells, and let's make each shell one light year thick, why not,' said Lorenzini. 'A finite number of stars will be in each shell, and assuming the universe to be homogeneous on the largest scale, which is an assumption we can, I think, *make,* we can calculate how many stars there are in each shell, the bigger they get. So there would be four times as many stars in a second shell, approximately, and because the second shell is twice as far away, each star in it would appear four times dimmer than the first shell. It cancels, you see. It means that the greater number of stars at a distance equals the same brightness. Each shell produces the same amount of light, going out and out forever in this, our infinite universe.'

'Joe,' pleaded Broome.

'It should be light. God is light. Where is it all? Where is all the God? You can tell me, can't you, Rev?'

'Whatever you have done, said Broome, feeling an unpleasant sensation of fear inside his gut. 'God can forgive – but you must open your heart to him.'

'My heart's already open to God. It's always been open,' Lorenzini said. 'But it's always been too dark for God to see. It's always been night. That's why we crave light.' A dog-like bark: laughter, perhaps. 'I'm a bloody poet, Rev, and I didn't know it.'

'Joe. You have to call an ambulance. For Kay. For me. Joe, for *you*. We need to get away from here, back to civilisation, back to people who can help us.'

'Astronomers,' Lorenzini's voice came out of the dark, 'never really took Olbers' Paradox seriously. It's always been treated as a trivial matter. There are half a dozen possible explanations, none of them very convincing, and everyone just shrugs their shoulders and says, "*Meh*." Fucking meh! But if you think about it for twenty seconds you soon see it's the most important question in astronomy – in science – in life, Rev. Because it can be answered so very simply. The sky should be all light, and it's not. The light beams out from an infinity of stars, and it ought to rain down upon our world too. But it doesn't. There can only be one explanation: because something is stopping it. And when you start thinking like that it gets inside your head, Rev. God is pouring light into his creation, and something is devouring almost all that light before it can reach us. You can call it dark matter or dark energy, but it doesn't matter what you call it. It sucks light.'

Then Broome heard the door opening, and he tried to rouse himself, but he was too exhausted and hot and dehydrated. Lorenzini went out singing to himself:

'It's always been night. It's why we crave light.
It's always been night. It's why we crave light.
It's always been night. it's why we crave light.'

Broome sat in the dark space and contemplated darkness. At one point thirst drove him to explore the place, crawling on all fours this time. His hands touched hard, warm, dry, something deliquescent and foul-smelling, something soft – a fold of cloth, large as a pillowcase. He couldn't work out what it was. Then he lay with his back to the wall and slept. He figured that the shaking of the metal walls would wake him, the next time Lorenzini opened the door.

It didn't happen. The next thing he knew was Lorenzini's voice, floating down as he drifted back up to consciousness to meet it.

'Adam and Eve were blind,' Lorenzini was saying. 'That's what the serpent tempted them with: sight. You can't know anything when you can't even see. Eat this therapeutic fruit and your eyes will open and the light will get in. And they did see – they saw that they were naked.'

His back hurt where he had banged it. Or cut it. Or he wasn't sure what he had done – pulled something, snapped something. He was not a young man.

'Are you naked, Joe?' he asked, in a creaky voice. 'Right now?'

'God blessed you, Rev, when he took your sight away. Kay couldn't see past the outside of me – the old skin, the ugly face, the grotesqueness of my flesh. She saw that, and that's where her seeing stopped – blocked, by her preconceptions. Pride. Dark energy. My gift to her, and she didn't really appreciate it. But she couldn't. She was trapped in the dark matter of her body.'

Broome started weeping, but quietly. 'What have you done, old friend? What have you done?'

'I have to apologise to you, Rev. When the darkness wraps me, I see now that the balance will mean that sight will return to you. I feel like I'm robbing you of God's gift of blindness. But dark energy means that the cosmos is expanding further and faster, leaving the big bang further and further behind. Dark energy is released when dark matter converts photons from light to – whatever it is that dark matter has, whatever gives it mass. Not gravitons. Satanitons, I guess.'

'Joe,' Broome said, but whatever else he was going to say was interrupted by a huge clattering noise, like a huge piece of metal being snapped abruptly in half, or like a giant kicking the door with steel-tipped toe-caps. The sound was so loud Broome's ears sang like Maria Callas hitting the high note in the Queen of the Night's aria.

He lay there for a long time, in the heat and the flawless darkness. It occurred to him that there might be a phone on Lorenzini's body; but it didn't seem likely, and even if there were he really wasn't sure he was going to be able to use it.

He explored the cut in his back with his fingers. Lying down was making him feel ill, so he pulled himself up. A dark element in a dark cell in a dark universe. One tiny portion of one tenth of one twentieth of one whole.

Suddenly he saw.

Afterword:

Olbers' Paradox

Prof. Sarah Bridle

Jodrell Bank

WHEN I CAME OUT of my interview for an undergraduate place at Cambridge University, I had many things on my mind. That I would be writing a story afterword on the same topic as my interview question 20 years later was not one of them.

Adam Roberts manages to get across a wide range of ideas in this entertaining tale. Yet the strangest of them – that we don't know what most of our universe is made of – is actually true. When we look at how fast galaxies and stars are moving, we see that there must be more gravity pulling them around than expected from the number of stars we can see alone. We can measure the amount of ordinary matter in the Universe by looking at the amount of hydrogen and helium produced in the minutes after the Big Bang, and this tells us that this extra gravity cannot come from the same ordinary material that we are made of, but must be from some new type of matter that has not yet been observed.

We believe the Big Bang set the Universe off expanding quickly, and we expect that the gravity of matter (including dark matter) will act to slow down this expansion – just like the gravity of the earth slows down the rise of an apple thrown into the air. When astronomers looked at the sky to measure this deceleration, they found a surprising result. The result was as surprising as throwing an apple into the air and seeing it take off into space. Astronomers found that the universe appears to be expanding faster and faster. Why this is happening is a complete mystery, and we call the source of this acceleration 'dark energy', although this is really just shorthand for 'the thing that is causing the acceleration'.

There are many as-yet unobserved particles predicted by particle physics theory that could make up the dark matter. The most popular candidates are termed weakly interacting massive particles (WIMPs), which includes a range of particles produced in supersymmetry. Searches are ongoing at the LHC to detect supersymmetry, which may shed light on the dark matter.

As Adam's story indicates, hints might be seen at the Large Hadron Collider at CERN that would give us clues as to which of these particles might make up this Dark Matter. Even stranger than Dark Matter, though, is Dark Energy – which is a whole different story. This isn't a mass that we can't see, but rather a mysterious energy or force in the cosmos that seems to be propelling the universe apart at a faster and faster rate. In 1998 and 1999, two separate research groups[36] discovered evidence that suggested the universe is not just flying apart, it's accelerating! So far, there are no good theories about what the dark energy is. The best theory is that space is full of particles appearing and disappearing, which creates a vacuum energy. However, this theory predicts a lot more dark energy than we actually observe, by a staggering factor: 1 followed by over 100 zeros, known as the vacuum catastrophe. Indeed, many cosmologists wonder whether a theory with two such large missing pieces could be correct, and whether we are instead on the verge of a paradigm shift in which we discover that Einstein's General Relativity is itself incorrect.

In Adam's story, the stir-crazy astronomer, Lorenzini poses the centuries-old question of why the sky is dark at night. This is often referred to as 'Olbers' Paradox', after the German astronomer Henrich Olbers (1758-1840), who also discovered a number of asteroids and first proposed that the asteroid belt was a destroyed planet. The 'Paradox' that he's most famous for, however, was actually proposed by others centuries before him, first by the English astronomer, Thomas Digges, then later by Johannes Kepler. The key mathematical result that underlies the paradox can be described as follows: the amount of light from the sun obscured, for example, by a pinhead at arm's length would be the same whether the sun were nearer or further away from us (unless the sun is so far away that it appears smaller than the pinhead). How can this be true? If the sun were closer then we would receive more light from the sun, but also the sun would appear bigger on the sky, so the larger amount of light is spread over this larger area, and the two factors cancel. If the universe were infinitely big and infinitely old then in every single direction there would be a star, and so the night sky would be as bright as the sun.

It was postulated at the time that there may be intervening material such as cold dust which absorbs the light. But this is not a satisfactory solution because the dust would heat up due to the illumination by the stars and eventually get so hot that it would itself emit light. This same criticism can be leveled at Joe's solution to Olbers' Paradox in the story, which is that 'dark matter swallowed the light.' Furthermore, the defining property of dark matter is that it does not emit or absorb (swallow) light.

The solution to Olbers' Paradox is simply that the universe has a finite age (13.8 billion years, as it turns out). When we look at distant stars we are also looking back in time, because of the time it takes for light to travel to us. The first stars formed when the Universe was a few hundred million years old, and so it is not the case that there is either an infinite scattering of stars proceeding into the distance, with an infinite amount of time for all of those star's light to reach us.[37]

Bizarrely, part of this solution (the idea that some starlight might not yet have reached us) was first proposed, not by a physicist, but by a short story writer, one of the fathers of the modern short story in fact: Edgar Allan Poe. In his 1848 essay, *Eureka*, Poe observes:

> 'Were the succession of stars endless, then the background of the sky would present us a uniform luminosity, like that displayed by the Galaxy – since there could be absolutely no point, in all that background, at which would not exist a star. The only mode, therefore, in which, under such a state of affairs, we could comprehend the voids which our telescopes find in innumerable directions, would be by supposing the distance of the invisible background so immense that no ray from it has yet been able to reach us at all.'

In Adam's story, Lorenzini claims that there is a second version of Paradox arising from the Big Bang theory: 'All our lines-of-sight should lead us back to the Big Bang. And that was presumably very, very bright.' He is right that we can actually see beyond the first stars, but we can't see quite as far as the Big Bang itself. What we see is the light leftover from the early fireball phase of the universe, when it was so hot that atoms were split into electrons and protons. The light scatters off free electrons like light scatters off water droplets in a cloud. Just like we see the under surface when we look up at a cloud, we can see the 'last scattering surface' of the light in the Universe, and this is called the Cosmic Microwave Background radiation, which was formed when the Universe was about 400,000 years old. The temperature of the Universe at that time was about the same as the temperature of the surface of the sun, so the light would have been visible to human eyes, had we been around at that time. However, because the universe is expanding, this light has been stretched out to longer wavelengths than our eyes can see, via the Doppler effect – the same effect that makes a police car siren sound higher-pitched when it's coming towards us. Astronomers call this Doppler effect of stars moving away from us 'redshift', because it shifts the wavelength in the direction of the red (and infrared) end of the spectrum, rather than the blue (or ultraviolet) end. So the Cosmic Microwave Background radiation has its wavelength stretched, for us, to that of the radiation you get in a microwave oven. So, when we look up at the sky, our eyes cannot see the light from the Big Bang.

In conclusion, Adam's story gives a thought-provoking tour of interesting questions in cosmology, even though Olbers' Paradox has already been solved, and the solution is not related to the dark matter or dark energy. Nowadays when I give public talks I meet a common misconception that dark matter and dark energy 'swallow' light – members of the audience almost always ask whether black holes could be responsible. I blame the nomenclature of the cosmologists. I think we should take up the suggestion of one audience member and rename them 'transparent matter' and 'transparent energy'. Or ideally the next Einstein will soon put both mysterious substances out of business with a replacement for General Relativity itself.

Notes

36. The High-Z Supernova Search Team, lead by Adam G Reiss and Brian P Schmidt, published observations of Type Ia ('one-A') supernovae. In 1999, the Supernova Cosmology Project, lead by Saul Perlmutter, followed by suggesting that the expansion of the universe is accelerating. All three received the 2011 Nobel Prize for Physics.

37. According to Edward Robert Harrison's *Darkness at Night: A Riddle of the Universe* (Havard University Press, 1987), the first formulation of this solution was in a little known paper by William Thomson (Lord Kelvin). For a key extract from this paper, see Harrison (1987), pp.227-28.

Inertia

Anneliese Mackintosh

'THE SPEED OF LIGHT is constant,' he says. 'It doesn't matter whether you're dying of cancer on the moon, the Andromeda Galaxy, or the Horologium Supercluster. The speed of light is always two hundred and ninety-nine million, seven hundred and ninety-two thousand, four hundred and fifty-eight metres per second.'

My dad is dying of cancer, and I am in his bed. My dad is dying of cancer, and I am nodding at appropriate intervals.

Albert Einstein is a sixteen-year-old boy. He has not quite grown into his facial features yet. His hairline is high for his age, his skull is large and angular, and his ears are low. He has a straight back and rounded shoulders, slightly hunched from reading. When he has nothing else to do, Albert Einstein likes to build houses of cards. He puts an ace of diamonds at the top.

Albert Einstein is a clever boy. He gets good marks in mathematics, and thinks carefully about what he says before he says it. Sometimes, minutes or even hours will pass before he replies to questions. Sometimes, he decides it's not worth replying at all.

Albert Einstein is a naughty boy. He hates going to school and sneaks looks at dirty pictures. He stays out late, kissing girls with Italian accents under bridges. He asks the girls if they would like to do unspeakable things to him.

Albert Einstein is a completely average teenage boy. Albert Einstein is a genius.

'You need to stop hugging me now,' my dad says into my arms. 'I'm on fire.'

I roll to the far side of the mattress and think icy thoughts. 'Has the morphine kicked in yet?' My dad is experiencing tremendous pain in his spine.

'I feel like the Switchback of Notre-Dame.'

'Sounds like a roller coaster,' I reply. 'Tell me about Special Relativity, Dad.'

My father breathes a little life out of his lungs, and begins.

When Albert Einstein was five years old, his father gave him a compass. Little Einstein was in bed with a cold, and as the compass was placed in his hands, he sneezed. 'Danke, Vati,' he said, and his daddy left the room.

The compass had a golden casing, and its needles were slender and red. Wee Albert placed the compass in his lap, and watched the needle swing towards the chair in the corner of his room.

Something is happening here which I cannot see, he thought.

'Let me tell you about inertia,' my dad says, as he continues to die. 'It comes from the Latin *iners*, meaning sluggish, stagnant, helpless, or weak.'

I close my eyes.

'Scientifically speaking, the law of inertia looks at resistance. It's the resistance of an object to change, whether that object is still, or moving with constant velocity. An object will only change velocity when an external force is applied. Are you with me?'

'How many weeks did the doctor say you have left?' I ask.

'Ten.' He stares into space. 'Roughly ten weeks left.'

Like all sixteen-year-old boys, Albert Einstein wants to be the best. He wants to win the most prizes, lift the biggest weights, grow the biggest penis, and run faster than anyone has run in the history of the world, ever.

Yes, he wants to run as fast as a train. He wants to run alongside the train, at exactly the same speed as the train, so it looks like the train is not moving at all.

No, he wants to run faster than that.

'When we start thinking about two or more moving bodies,' my father explains, 'things get more complicated. Bodies might be moving with uniform velocities in respect to each other, or with non-uniform velocities. To distinguish between them, we use inertial versus non-inertial reference frames.'

There are sheep in the field outside. When I was five, my dad learnt how to inject sheep's eyelids. He spent two weeks at agricultural college, and he bought five sheep. Those sheep kept us in mutton for three years.

'In my opinion,' I tell my father, 'being resistant to change implies great strength. Not weakness. Inertia should be called endurance, forbearance, stoicism.'

'In the eighteenth century,' he replies, 'people believed that the resistance of bodies to acceleration was not a passive feature of bodies, but an active one.'

We both lie perfectly still.

Albert Einstein likes to conduct thought experiments. Most of his experiments involve imagining what it would be like to run really, really fast. He dreams of winning the one-hundred metre race at Sports Day, being called up in front of the whole school, climbing onto a platform and receiving a trophy.

He dreams of going so fast his body is nothing but a blur on the horizon.

He dreams of going as fast as the speed of light.

'Remember that place we used to get sheep nuts?' I ask my father. 'That farm? We used to drive out there, listening to Radio 4. I would look out of the window, up at the sky, and think about the enormity of the world. That journey used to take hours.'

'It was a five minute drive,' my father says. 'You know about time dilation?'

In this thought experiment, that girl he fancies – Marie or Helga or Clare – she is watching him run, cheering him on. She will take him in her arms and fuck him tonight. His mathematics teacher, Frau Grüner, she is there too. She will award him that scholarship and give him ten out of ten. Everyone is there, watching him ping through the air, dart over fields and oceans, stars and galaxies.

'I'm getting ahead of myself,' my father tells me, after a long pause, a pause in which I'm sure he just fell asleep, 'but if you're quick enough, time will begin to slow down for you.'

I blink, and imagine my blink to last a lifetime.

'If I was to fly into space right now – really, really fast – and you were to keep lying here, just like this, then I could return after you'd lived for ten years, and I might only be one second older.'

I look at the lines on the backs of my hands. 'I would be forty-one,' I say, now examining my life-lines. 'I would have wrinkles and grey in my hair.'

'And I would be dying,' he says, 'no more and no less than I am right now.'

This is what it means to be alive.

Synapses gushing.

Mind expanding.

Albert Einstein is a fricking fast, fricking sexy genius.

'If I kept travelling fast enough,' my father says, 'I could really make use of the ten weeks I have left. I could zoom off in a rocket, and come back for important events. I could give you away on your wedding day, do the two-times-table with my first grandchild, attend the graduation of my second. I could sing happy birthday at your fiftieth, and shake your hand when you retire.'

'You could hold me the day that I die.'

'If I had a spaceship full of morphine,' my dad says, 'I feel like anything would be possible.'

Something unbelievable is happening; Albert Einstein is about to become champion of the universe. He is going to ride on top of a beam of light. He will balance on it like a surfboard, with his arms out and his head high. He will look down at his adoring Helga, back on Planet Earth. He will be der Gewinner. He will have a hypergiant penis.

'You know what?' Dad says. 'If Einstein had travelled fast enough when he was a young man, then he might still be alive, somewhere out there in the non-existent ether. I could go and meet him for a cup of tea.'

I watch two sheep, nuzzling by the hedgerow. 'What would you say to Einstein, if you could meet him now?'

My dad thinks about this for a long time. So long he manages to squeeze in another nap. 'I would say, "Hello",' he says eventually. 'Though I would probably speak in German, so it would be "Hallo".'

'What else?'

'I'd ask him to talk me through Special Relativity, so I could explain it to you better than this. So I didn't have to make up the juicy details.'

'What else?'

'I'd ask him the one thing he was most proud of, and his

biggest regret.' As the words come out, my father's breath slackens. 'I'd ask him his happiest memory. That's it, yes. His happiest memory.'

'I love you, Dad,' I say, inhaling sharply. I will breathe for both of us from now on.

Albert Einstein takes one giant leap–
But something is wrong.
The beam of light is escaping.
Albert Einstein is stuck in second place.

'It's impossible for any object to travel at the speed of light,' my father tells me. 'I'm afraid you just can't do it. The speed of light will always win. Think about $E=mc^2$. Energy is equivalent to mass times the speed of light squared. An object gains mass as it accelerates.'

He begins to talk, faster and faster. I think about the coldness of the pillow on my cheek. I am close to my father, but I don't always understand him. Trying to follow his train of thought is like trying to catch up with a beam of light.

'When I look at the walls, everything turns purple,' he says. 'I wonder how much fluid is left in my spinal cord.'

'I'll make some dinner,' I say. 'Beetroot soup?'

'Crackers,' he says. 'I think we should have cheese and crackers to celebrate.'

I don't ask what we're celebrating, but I think I know.

Albert Einstein is turning seventeen. He's having a big party. There will be sausages, cheese, fruit, and tea.

Alas, he is another year older and life is passing him by. He is just a human being, made up of blood and tissue and bone. Nothing special.

It seems like my father's morphine is wearing off. He's grinding his teeth and the clock is starting to tick more slowly for him.

'I'll get the crackers,' I say.

Albert Einstein spends the rest of his life doing the stuff that people do. He writes essays and plays violin concertos and comes up with groundbreaking theories. He is given certificates and causes controversy and finds true love. He takes a shit every morning and holds his children at night and he is dying, every day he is here he is dying, and soon he will be dead.

'Wait,' pleads my father. 'I'm not ready. Let's lie here a moment longer.'

'Okay,' I say, wondering if he wants to be buried or burned. 'But we should eat before it gets dark.'

I reach out to him, across the mattress. There are many, many cubic millimetres of bed linen between us. Each cubic millimetre holds billions and billions of molecular structures, tiny atoms, which, when viewed close-up, resemble galaxies, just like our own.

'I wish I'd achieved more,' my dad says quietly, tears in his eyes.

'You've achieved the miracle of life,' I reply. 'That is the most remarkable thing in the universe.'

Outside, the sheep move slowly up the hill, towards shelter. Inside, my fingers search, probe through endless galaxies, until finally, I find my father's hand. We touch palms – life-lines locked together – and we stay like this, still and helpless, strong and inert, flying through time and space, dying, both of us dying, but never as fast as the speed of light.

Afterword:

Chasing a Beam of Light

Prof. Michela Massimi

University of Edinburgh

ANNELIESE'S STORY IS CENTERED around one of the most baffling aspects of relativity theory. The so-called 'light principle' is a postulate of Einstein's Special Relativity, which says that the speed of light in a vacuum (i.e. not through media) has a constant value *c* (just under 300,000 km/sec) for all inertial reference frames. In other words, the velocity of light has a constant value for any observer, who is either at rest or is moving with uniform velocity (that's what 'inertial reference frame' means: any reference frame that follows Newton's first law). This principle is a cornerstone of Einstein's Special Relativity. It is an axiom (not a theorem) of Special Relativity.[38] Important consequences follow from it: for example, *time dilation* (i.e. the rate at which clocks tick varies with the speed of the reference frame) and *length contraction* (i.e. the length of objects is relativistically affected by the speed of the reference frame). Thus, to understand how Special Relativity has changed our deeply entrenched intuitions about the nature of space and time, we ought to understand the light principle.

In 1905, at the age of 26, Einstein published a series of papers, which turned the world of physics upside down. And to understand why, let us take a brief look at two main physical theories before Einstein: Newtonian mechanics and Maxwell's electromagnetism. Newtonian mechanics tells us that if we want to describe the motion of a mass point, we can use a reference frame (roughly a system of three Cartesian coordinates x, y, z, with an origin O). Newtonian mechanics describes how the mass point changes its position over time in this Euclidean space as a result of moving with uniform motion (let us leave aside for now non-uniform motion, which falls outside the scope of 'inertial reference frames').

Thinking along these Newtonian lines, in our daily lives we are accustomed to thinking of velocities as relative. Imagine I throw a ball. Let us assume for a moment (just for the sake of what it is to be an inertial reference frame) that the ball moves with uniform velocity at say 7m/sec. If I run fast enough (i.e. 7 m/sec), I will be able to catch the ball. Or, imagine I am sitting on a train that is moving with uniform velocity of 50 km/hour, and my friend Mary in her red sports car is driving along the road (parallel to the train track) at a uniform velocity of 50 km/hour. I should be able to see and wave at Mary from the train. Mary will appear to me at rest in her sports car. But from the point of view of someone at the train station, Mary will appear to be moving away from the train station at 50 km/hour. Scenarios like these are ubiquitous in our everyday lives because we are accustomed to thinking about the world as a Newtonian world.

But now consider some profoundly counterintuitive consequences that follow from this Newtonian line of reasoning. Like the example of the ball, or of my friend Mary in her car, one might ask what the velocity of a beam of light would be if one were able to run fast and chase the beam of light. If the Newtonian line of reasoning were correct, the answer would seem to be 'zero'. If I were able to run fast enough, indeed close to 300.000 km/sec, the beam of light would appear to me at rest (just as my friend Mary seems to be at rest when I am sitting on the train and travelling down the track at the same speed as Mary's car).

But if this were indeed the case, it would be bad news for another very successful scientific theory: Maxwell's electromagnetism. For Maxwell's theory tells us that the speed of light is a constant *c*. Maxwell's theory tells us that the speed of light has an invariant value (incidentally, a value that was experimentally measured at the end of the nineteenth century to be around 300,000 km/sec). Either the speed of light is constant (as Maxwell's theory tells us), or, it is not (as Newtonian mechanics seems to suggest). How to resolve this contradiction? In his *Autobiographical Notes*, Einstein recounts how this thought occurred to him at the age of sixteen, by imagining what would happen if he were able to chase a beam of light:

> 'After ten years of reflection, such a principle resulted from a paradox upon which I had already hit at the age of sixteen: If I pursue a beam of light with the velocity *c* (velocity of light in vacuum), I should observe such a beam of light as a spatially oscillatory electromagnetic field at rest. However, there seems to be no such thing, whether on the basis of experience or according to Maxwell's equations. From the very beginning it appeared to me intuitively clear that, judged from the standpoint of such an observer, everything would have to happen according to the same laws as for an observer who, relative to the earth, was at rest. For how otherwise should the first observer know, that is, be able to determine, that he is in a state of uniform motion? One sees in this paradox the germ of the Special Relativity theory is already contained.'[39]

This misleadingly simple thought experiment brings to the fore a profound truth. The speed of light *c* as it appears in Maxwell's equations would no longer be constant. It would become relative to inertial reference frames (i.e. it would be 300,000 km/sec in a inertial reference frame that is at rest in what at the time was assumed to be an all-pervasive ether as the medium of light; but it would be less than 300,000 km/sec in any other inertial reference frame moving away with uniform velocity).[40]

But, surely, we want our laws of nature not to be relative.

One would expect Maxwell's equations for the electromagnetic field to be invariant, no matter which inertial reference frame is considered. Einstein's solution to the conundrum was to retain Maxwell's equations and ditch Newtonian mechanics (and the notions of space and time associated with it). One will never be able to catch a beam of light, no matter how fast one can possibly run in the thought experiment. The speed of light is a constant: it does not change by changing inertial reference frame. It will never get to 0 by riding a beam of light. This profoundly counterintuitive conclusion of Einstein's famous thought experiment is behind the light principle. And on it, Einstein laid the foundations of Special Relativity: the relativity of simultaneity, time dilation, and length contraction, all follow from the light principle.

But, obviously, life is a lot more messy and complicated than in any (fantastically simple) thought experiment. The beauty of Anneliese's story is to bring this thought experiment to life by imagining a young, adolescent, slightly naughty Einstein day-dreaming about beams of light and whatnot. And the possibility of time dilation becomes a wish in a hospital bed, in the vain hope that clocks might stop ticking and death evaded. When I first read Anneliese's draft in February 2014, I would have never imagined that a year and half later, I would have found myself in a hospital bed, witnessing my own sixty-seven-year-old father dying of cancer. In those days, I kept thinking about Anneliese's story, and like her character, I too wished that time could be dilated and death evaded. When Michele Besso, one of Einstein's best friends, passed away, in a letter sent to Besso's family Einstein famously said: 'Now he has departed from this strange world a little ahead of me. That means nothing. People like us, who believe in physics, know that the distinction between past, present, and future is only an illusion, although a persisting one'.[41] Anneliese's story is a hymn to this temporal illusion – persisting, fleeting, and painful as it can be, even for those who believe in physics.

Notes

38. In other words, if we follow Einstein's 1905 formulation of Special Relativity as an axiomatised kinematics, the light principle features as an axiom (although it is obviously possible to deduce the light principle from Maxwell's electrodynamics and the so-called principle of relativity).

39. A. Einstein 'Autobiographical Notes', in *Albert Einstein. Philosopher-Scientist*, ed. P. A. Schilpp (New York: Tudor, 1951), pp.52-53. The story behind this thought experiment is very interesting. The historian of physics John Norton has argued that at the age of sixteen, in secondary school (gymnasium) at the Canton of Aargau, Einstein could not have possibly known Maxwell's equations. Instead, Norton suggests that Einsten's later recollection of this thought experiment in the *Autobiographical Notes* might have been prompted by Einstein's reading of Max Wertheimer's book *Productive Thinking*. Wertheimer was a psychologist; his book was published posthumously in 1945 and it contains an early interview with a young Einstein, where the sixteen-year-old boy's musing about the beam of light could be found. According to Norton, the *prima facie* misleading simplicity of this thought experiment was probably 'coaxed out of Einstein by an eager Wertheimer. His hours and hours of interrogation … were those of a psychologist with a definite view. He needed a simple story he could relate to non-technical readers', Norton (2016) 'How Einstein Did Not Discover', pre-print in www.pitt.edu/~jdnorton. Norton makes a persuasive case that far from being simple, the thought experiment is in fact hard to make sense of if one does not take into account the long years leading up to the 1905 paper, years in which the young Einstein tried to develop an emission theory of light without success: 'it is not at all clear how this thought experiment works. In the dominant theories of the late nineteenth century, light propagates as a wave in a medium, the luminiferous ether. It was an entirely uncontroversial result in the theory that, in a frame of reference that moved with the light, the wave would be static. There is no reason for us to be puzzled. We do not see frozen light since we

are not moving at the speed of light through the ether' Norton (ibid., p. 13). For readers interested in the technical details of this story and how Einstein's long years of work on electrodynamics led him to the light principle, please see John Norton 'Einstein's Investigations of Galilean Covariant Electrodynamics Prior to 1905', *Arch. Hist. Exact Sci.* 59 2004, 45–105. Pre-print available for download at www.pitt.edu/~jdnorton.

40. To be precise, as John Norton explains it, Einstein's thought experiment does not provide a compelling objection against ether-based theories of electrodynamics. Instead, it is meant to provide an objection against emission theories of light prior to Einstein's 1905 Special Relativity, because emission theories allowed light to slow and eventually freeze for an observer chasing it. Emission theories are incompatible with Maxwell's theory, which forbids frozen waves of perpendicular electric and magnetic fields in a vacuum. For details see John Norton (2013) 'Chasing the light: Einstein's most famous thought experiment'. In J. R. Brown, M. Frappier, L. Meynell (eds.) *Thought Experiments in Philosophy, Science and the Arts* (New York: Routledge 2013), 123-140. Pre-print available for download at www.pitt.edu/~jdnorton.

41. P. Speziali (ed.) *Albert Einstein, Michele Besso. Correspondence 1903-1955* (Paris: Hermann, 1972). Letter 215. 'Einstein au Fils et la Soeur de Besso' (Princeton, 21. III. 55), p. 538.

About the Authors

Sandra Alland is a Scotland-based writer and artist who has published and presented throughout the UK, Europe and North America. Recent highlights include *subTerrain, Feral Feminisms, Gutter*, English PEN, Tate Modern, Schwules Berlin and Comma's forthcoming *The Mirror in the Mirror.* Sandra has published three books of poetry and a chapbook of short fiction, and is co-editor of *Stairs and Whispers: D/deaf and Disabled Poets Write Back* (Nine Arches, 2017).

Annie Clarkson is a writer and social worker. Previous stories have been published in Comma anthologies, *Brace* (2008), *Litmus* (2011), *Lemistry* (2012), and *Spindles* (2015) as well as various anthologies and magazines.

Marie Louise Cookson writes and performs short fiction, monologues and scripts. Her play, *Get The Confidence, Get The Love* was produced as part of Shortworks, also at Manchester's Contact Theatre. She has collaborated with the film and video collective, Institute Zoom on their fantastical time adventure serial, *Universal Ear.* She also writes for the blog, *Well Fan My Brow & Shut My Mouth.*

Claire Dean's short stories have been widely published in anthologies, and two chapbooks by Nightjar Press. Her first collection *The Museum of Shadows and Reflections* was released by Papaveria Press this year.

Zoe Gilbert's short stories have appeared in anthologies and journals in the UK and internationally. She is currently working on her first collection of stories inspired by folk tales, and is studying for a PhD in creative writing at the University of Chichester. Her story 'Fishskin, Hareskin' won the Costa Short Story Award 2014.

Andy Hedgecock has written reviews, essays and non-fiction for 30 years for the likes of *The Morning Star*, *The Spectator, Time Out, Penguin City Guides,* and a number of SF magazines and academic journals. He is a writer, researcher and former co-editor of *Interzone*, Britain's leading SF magazine.

Robin Ince is a comedian, actor and writer, perhaps best known for presenting the BBC radio show *The Infinite Monkey Cage* with physicist Brian Cox. His touring shows have included *Bleeding Heart Liberal* and *Happiness Through Science.* In 2014 he teamed up with Johnny Mains to co-edit *Dead Funny*, an anthology of horror stories by comedians.

Annie Kirby is a short story writer, novelist and writing tutor. Her stories have appeared in various anthologies, including Comma's *Bracket, Bio-Punk,* and *Beta-Life*. Her Asham Award winning story 'The Wing' was published in *Don't Know A Good Thing* (Bloomsbury) and adapted for audio download by Spoken Ink. Her stories have been selected for new writer showcases including BBC Radio 4's *Writers to Watch* and the *Portsmouth 2012 Bookfest* anthology. She lives in Portsmouth and has recently completed her first novel.

Anneliese Mackintosh's short story collection, *Any Other Mouth*, published by Freight, won The Green Carnation Prize in 2014, and was shortlisted for the Edge Hill Prize, Saltire Society's First Book Award, and the Saboteur Award, and was a 'Book Of The Year' in T*he Herald, The Scotsman, Civilian,* and *The List Magazine,* as well as one of *The Guardian* readers' 'Top Ten Books of 2014'. Anneliese's first novel, *So Happy It Hurts*, is published by Jonathan Cape this year.

Adam Marek is the award-winning author of two short story collections: *Instruction Manual for Swallowing* and *The Stone Thrower*. He won the 2011 Arts Foundation Short Story Fellowship, and was shortlisted for the inaugural *Sunday Times* EFG Short Story Award.

Adam Roberts is an academic, critic and novelist. He has a PhD from Cambridge University on Robert Browning and the Classics. His novel *Jack Glass* (Gollancz 2013) won the John Campbell and the BSFA Awards. His latest novel is *The Thing Itself* (Gollancz, 2015), an SF novelisation of Kant's *Critique of Pure Reason* via John Carpenter's *The Thing*.

Sarah Schofield's prizes include the Writers Inc Short Story Competition and the Calderdale Short Story Competition. She was shortlisted for the Bridport Prize in 2010 and was runner up in *The Guardian* Travel Writing Competition. She has contributed to four previous Comma anthologies.

Ian Watson taught in Tanzania, Japan, and Birmingham's School of History of Art before becoming a full-time writer in 1976 after the success of his first SF novel, *The Embedding*. 40 books followed; he is translated into 17 languages. 2014 saw *The Best of Ian Watson* from PS Publishing. His thirteenth story collection, *The 1000 Year Reich*, debuts from NewCon Press, Easter 2016. Ten months eyeball to eyeball with Stanley Kubrick resulted in screen credit for *A.I. Artificial Intelligence* (2001), filmed by Steven Spielberg.

Margaret Wilkinson is a short story, stage and radio writer, as well as a senior lecturer in creative writing at Newcastle University. Her most recent radio play for BBC Radio 4, *Nocturne*, merged mothers, daughters, and Chopin. In all her writing she is deeply interested in, and inspired by, character and the sound, tone, and rhythm of voice.

About the Scientists & Philosophers

Dr. **Rob Appleby** is a Reader of physics at the University of Manchester, a member of the Accelerator Physics group and of the Cockcroft Institute. His primary research interest is the physics of particle accelerators, including the motion of particles and new accelerator concepts, and he has worked for many years on the Large Hadron Collider and particle accelerators for cancer treatment. He holds a PhD in theoretical physics from the University of Manchester and a Masters degree in theoretical physics from the University of York. He has previously been a consultant on Comma's *Litmus* and *When It Changed* anthologies.

Professor **Stewart Boogert** is Deputy Director of the John Adams Institute for Accelerator Science, a joint venture between the universities of Oxford, Imperial College, and Royal Holloway College. His research spans a portfolio of accelerator research activities at Royal Holloway. Having received his D. Phil in particle physics from Oxford, St. Catherines College, in 2002, he was a Post Doctoral Research Assistant in particle physics, specifically the Linear Collider at University College London. He has been an academic staff member at Royal Holloway in accelerator physics since 2005.

Sarah Bridle is Professor of Astrophysics in the Extragalactic Astronomy and Cosmology research group of the Jodrell Bank Centre for Astrophysics, in the School of Physics and Astronomy, at the University of Manchester.

Seth Bullock studied cognitive science and gained a PhD in evolutionary simulation modelling at the University of Sussex, before working in Berlin and at the University of Leeds, where he founded the Biosystems Research Group. He has since been Professor of Computer Science at the University of Southampton (where he was Director of the Institute for Complex Systems Simulation), and is now the Toshiba Chair in Data Science and Simulation at the University of Bristol.

Steven French is Professor of the Philosophy of Science at the University of Leeds where his research interests include the philosophy of science and the history and philosophy of modern physics. He has previously published a book on structural realism (*The Structure of the World: Metaphysics and Representation*; OUP 2014) and is currently working on a series of papers on monism, dispositionalism and the relationship between science and metaphysics in general. He is Co-Editor-in-Chief (with Michela Massimi) of *The British Journal for the Philosophy of Science* (http://bjps.oxfordjournals.org/), and also Editor-in-Chief of the Palgrave-Macmillan series, *New Directions in Philosophy of Science.*

Roman Frigg is Professor of Philosophy at the London School of Economics, and Director of the Centre for Natural and Social Science (CPNSS). He holds a PhD in Philosophy from the University of London and masters degrees both in theoretical physics and philosophy from the University of Basel, Switzerland. His main research interests are in general philosophy of science and philosophy of physics.

Frank Jackson is an Australian philosopher, currently Distinguished Professor and former Director of the Research School of Social Sciences at Australian National University. He was also a regular visiting professor of philosophy at Princeton University from 2007 through 2014. His research focuses primarily on philosophy of mind, epistemology, metaphysics, and meta-ethics. In philosophy of mind, Jackson is perhaps best known for the Knowledge Argument against physicalism – the view that the universe is entirely physical (i.e., the kinds of entities postulated in physics). He is the inventor of the 'Mary's Room' thought experiment.

Ana Jofre received her PhD in Physics from the University of Toronto, did Post-doctoral work at the National Institute of Standards and Technology in Gaithersburg MD, and taught at the University of North Carolina in Charlotte for six years before transitioning her career towards the arts, and completing an MFA at OCAD University in Toronto. She then held research fellowships at in the visual analytics lab at OCAD University, and in Culture Analytics at the Institute for Pure and Applied Mathematics (IPAM) at UCLA. Currently, she is a professor of creative arts and technology at SUNY Polytechnic Institute in Utica, NY.

Michela Massimi is Professor of Philosophy of Science in the Dept. of Philosophy, at the University of Edinburgh. She received her PhD at the London School of Economics (2002), under the supervision of Michael Redhead and Carl Hoefer. She was a Junior Research Fellow at Girton College, University of Cambridge (2002-2005); and Visiting Professor in the HPS Dept., University of Pittsburgh (autumn 2009). From Oct. 2005 to June 2012, she taught in Dept. of Science and Technology Studies, University College London. Her research primarily focusses on philosophy of science, Kant, and the intersection between contemporary philosophical problems and historical and contemporary scientific practice.

Tara Shears is a particle physicist and Professor of Physics at the University of Liverpool. She has spent her career investigating the behaviour of fundamental particles and the forces holding them together, and has worked at experiments at CERN, the European centre for particle physics, and at the

Fermilab particle physics facility near Chicago, USA. Tara joined the LHCb experiment at CERN's Large Hadron Collider in 2004, where she currently works.

Glen Newey is Professor of Practical Philosophy at the University of Leiden. He previously worked in Brussels and until 2011 was Professor in the School of Politics, International Relations & Philosophy at Keele University. He is a prominent member of the 'Realist' school of political philosophers which also includes such figures as Bernard Williams, John N Gray and Raymong Guess. His books include *Hobbes and Leviathan* (Routledge, 2014) and *Toleration in Political Conflict* (CUP, 2013). He regularly blogs for the *London Review of Books*.

Jonathan Wolff is Professor of Philosophy and Dean of the Faculty of Arts and Humanities at University College London. He is a former secretary of the British Philosophical Association and the Aristotelian Society. He is the author of a critique of Robert Nozick's *Anarchy, State, and Utopia* called *Robert Nozick: Property, Justice and the Minimal State*, a short book on Karl Marx, *Why Read Marx Today?*, and *An Introduction to Political Philosophy*. He currently writes a monthly column for *The Guardian* and occasionally blogs at Brian Leiter's 'Leiter Reports' blog. In 2009, he presented a series about the NHS for BBC Radio 3's *The Essay*.

Special Thanks

Michela Masssimi and the editors would like to thank John Norton for his extremely helpful insights in the historical context of the 'Chasing a Beam of Light' thought experiment. The editors would also like to thank Manisha Lalloo and Sarah Barnes of the Institute of Physics for their unwavering support throughout the project. And finally, Ra Page would like to personally thank Prof. Harvey Brown, for his inspiration as an undergraduate.